Leveraging Blockchain Technology

Blockchain technology is a digital ledger system that allows for secure, transparent and tamper-proof transactions. It is essentially an often decentralized, distributed, peer-to-peer database that is maintained by a network of computers instead of a single entity, making it highly resistant to hacking and data breaches. By providing greater security, transparency and efficiency, blockchain technology can help to create a more equitable and sustainable world.

Blockchain technology has the potential to help mankind in various ways, some of which include but are not limited to:

- *Decentralization and Transparency*: Blockchain technology allows for decentralization of data and transactions, making them more transparent and accountable. This is particularly important in fields such as finance, where trust and transparency are critical.
- *Increased Security*: Blockchain technology is inherently secure due to its distributed nature, making it very difficult for hackers to compromise the system. This makes it an ideal solution for data and information storage, particularly in areas such as health and finance, where privacy and security are of utmost importance.
- *Faster Transactions*: Blockchain technology eliminates the need for intermediaries, reducing the time and cost associated with transactions. This makes it an ideal solution for international trade, remittances and other types of financial transactions, especially in parts of the world where a great number of individuals do not have access to basic banking services.
- *Immutable Record*: One of the fundamental attributes of blockchain is its immutability. Once data is added to the blockchain, it becomes nearly impossible to alter or delete. This feature ensures a tamper-resistant and reliable record of transactions, crucial for maintaining integrity in various industries, including supply chain management and legal documentation.
- *Smart Contracts*: Blockchain technology supports the implementation of smart contracts, which are self-executing contracts with the terms of the agreement directly written into code. This automation streamlines processes and reduces the risk of fraud, particularly in sectors like real estate and legal agreements.
- *Interoperability*: Blockchain's ability to facilitate interoperability allows different blockchain networks to communicate and share information seamlessly. This attribute is pivotal for creating a unified and interconnected ecosystem, especially as various industries adopt blockchain independently. Interoperability enhances efficiency, reduces redundancy and fosters collaboration across diverse sectors.

Leveraging Blockchain Technology: Governance, Risk, Compliance, Security, and Benevolent Use Cases discusses various governance, risk and control (GRC) and operational risk-related considerations in a comprehensive, yet non-technical, way to enable business leaders, managers and professionals to better understand and appreciate its various potential use cases. This book is also a must-read for leaders of non-profit organizations, allowing them to further democratize needs that we often take for granted in developed countries around the globe, such as access to basic telemedicine, identity management and banking services.

Security, Audit and Leadership Series

Series Editor: Dan Swanson, Dan Swanson and Associates, Ltd., Winnipeg, Manitoba, Canada

The **Security, Audit and Leadership Series** publishes leading-edge books on critical subjects facing security and audit executives as well as business leaders. Key topics addressed include Leadership, Cybersecurity, Security Leadership, Privacy, Strategic Risk Management, Auditing IT, Audit Management and Leadership

CyRM^SM: Mastering the Management of Cybersecurity
David X Martin

Why CISOs Fail (Second Edition)
Barak Engel

Riding the Wave: Applying Project Management Science in the Field of Emergency Management
Andrew Boyarsky

The Shortest Hour: An Applied Approach to Boardroom Governance of Cybersecurity
Lee Parrish

Global Audit Leadership: A Practical Approach to Leading a Global Internal Audit (GIA) Function in a Constantly Changing Internal and External Landscape
Audley L. Bell

Construction Audit: Building a Solid Foundation
Denise Cicchella

Continuous Auditing with AI in the Public Sector
Lourens J. Erasmus and Sezer Bozkuş Kahyaoğlu

Ironwill 360° Leadership: Moving Forward: Unlock Twelve Emerging Trends for Forward Thinking Leaders
Douglas P. Pflug

The CISO Playbook
Andres Andreu

Leveraging Blockchain Technology: Governance, Risk, Compliance, Security, and Benevolent Use Cases
Shaun Aghili

The Closing of the Auditor's Mind? How to Reverse the Erosion of Virtue, Trust, and Wisdom in Modern Auditing
David J. O'Regan

For more information about this series, please visit: https://www.routledge.com/Internal-Audit-and-IT-Audit/book-series/CRCINTAUDITA

Leveraging Blockchain Technology

Governance, Risk, Compliance, Security, and Benevolent Use Cases

Edited by
Shaun Aghili

CRC Press
Taylor & Francis Group
Boca Raton London New York

CRC Press is an imprint of the
Taylor & Francis Group, an **informa** business

Designed cover image: © Shutterstock

First edition published 2025
by CRC Press
2385 NW Executive Center Drive, Suite 320, Boca Raton FL 33431

and by CRC Press
4 Park Square, Milton Park, Abingdon, Oxon, OX14 4RN

CRC Press is an imprint of Taylor & Francis Group, LLC

© 2025 selection and editorial matter, Shaun Aghili; individual chapters, the contributors

ISBN: 978-1-032-61100-6 (hbk)
ISBN: 978-1-032-61105-1 (pbk)
ISBN: 978-1-003-46203-3 (ebk)

DOI: 10.1201/9781003462033

Typeset in Sabon
by SPi Technologies India Pvt Ltd (Straive)

Contents

Foreword

In an era of rapid technological advancement, the discourse surrounding blockchain technology has transcended its technical intricacies to become a focal point of managerial consideration. *Leveraging Blockchain Technology: Governance, Risk, Compliance, Security, and Benevolent Use Cases* meticulously delves into the multifaceted dimensions of blockchain's impact on governance, operational risks, compliance, as well as data privacy and security. Edited by Dr. Shaun Aghili, an award-winning Management professor and researcher in the field of Information Systems Security and Assurance Management at The Concordia University of Edmonton (CUE), this book represents the culmination of research conducted by a cohort of graduate students under his leadership during the 2023–2024 academic terms.

Blockchain technology, with its immutable digital ledger system, offers a paradigm shift in how we conceive security, transparency and efficiency in transactions. Through a lucid exploration of its benevolent applications, this book elucidates how blockchain transcends traditional boundaries to foster decentralization, transparency and heightened security. Dr. Aghili and his adept cohort ingeniously navigated the complexities of blockchain in a manner accessible to business leaders, managers and professionals, thereby empowering them to harness its transformative potential.

Furthermore, the book extends its gaze beyond the corporate realm, underscoring blockchain's capacity to democratize access to essential services globally. From self-governed identity management to inclusive banking solutions and fortified voting systems, the discussed use cases underscore blockchain's capacity to redress societal inequities and fortify democratic and humanitarian processes.

In essence, *Leveraging Blockchain Technology* transcends the realms of academia and industry, serving as a beacon for informed decision-making and societal progress. As Dean of Management at CUE, I commend Dr. Aghili and his esteemed Graduate student cohort for their scholarly rigor and visionary insights encapsulated within these pages. May this tome inspire stakeholders across sectors to embrace blockchain's transformative potential for the betterment of humanity.

Dr. Alison YacyShyn, Dean of Management
Concordia University of Edmonton (CUE)
Summer 2024

Preface

EXPANDING THE EXPLORATION OF BLOCKCHAIN TECHNOLOGY

In the ever-evolving landscape of information systems, blockchain technology has emerged as a transformative force reshaping the foundations of digital transactions. This preface chapter serves as a gateway to our latest book *Leveraging Blockchain Technology: Governance, Risk, Compliance, Security, and Benevolent Use Cases*. As we embark on this exploration, it is important to delve into the intricacies of blockchain technology, understand its evolution and unravel the motivations behind the creation of this second blockchain title.

At the core of blockchain technology lies a decentralized and distributed ledger system that revolutionizes the way we record digital transactions. The foundation comprises a chain of blocks, each encapsulating a list of transactions and linked chronologically forming an immutable and transparent chain. This *transparency, immutability* and *decentralization* are the pillars that elevate blockchain above all other traditional record-keeping systems.

Transparency ensures that transactions recorded on the blockchain are visible to all participants, fostering trust and accountability. Blockchain Immutability safeguards against fraudulent activities, thereby providing a secure and tamper-resistant environment. Decentralization eliminates the risks associated with a single controlling entity, distributing control across the network and enhancing overall security.

Distinguishing between public and private blockchains is also important. Operating in a permissionless and trustless environment, public blockchains offer openness and accessibility. In contrast, private blockchains restrict access, thereby providing controlled environments that prioritize privacy. In contrast to traditional ledgers, this decentralized nature fortifies the security of blockchain, making it resistant to manipulation and a reliable record-keeping system for the digital age.

The applications of blockchain technology are diverse and impactful. From facilitating secure financial transactions through cryptocurrencies to revolutionizing supply chain management and automating agreements with smart contracts, blockchain's influence extends across various industries. Identity management with its secure and tamper-resistant record-keeping and benevolent use cases further highlights the technology's versatility.

TRACING THE ROOTS OF THIS SECOND BOOK ON BLOCKCHAIN

The genesis of this book can be traced back to a collective effort by a dedicated graduate student team in 2020–2021. Our first comprehensive work *The Auditor's Guide to Blockchain Technology* (ISBN: 978-1-032-07825-0) paved the way for a more in-depth exploration of major blockchain domains. To further explain, in January 2021, a 55-member team comprised of 53 second-year students, two graduate assistants and me as the research project's advisor embarked on an extensive research project to create literature reviews on various aspects of blockchain technology.

Published in Winter 2023, *The Auditor's Guide* marked the culmination of our efforts and served as a launchpad for this second book. The subsequent research cohort sought to delve deeper into specific facets of blockchain technology, including data architecture, governance, operational risk management, privacy, security and taxation considerations.

A distinctive feature of this book is the emphasis on "Benevolent Blocks." This advocacy urges the use of blockchain for benevolent purposes, such as enhancing voting results, thereby further facilitating democracy and free choice, establishing a blockchain-driven system of self-governed identity, as well as providing various types of financial services for underserved demographics.

In particular, Chapters 6 and 12 emerge as interesting topics in this exploration. Chapter 6 delves into the intricate financial reporting and taxation aspects of digital assets, including cryptocurrencies and non-fungible tokens (NFTs). Chapter 12 explores the potential of blockchains to support and enhance Islamic finance principles, impacting over 1.8 billion Muslims worldwide.

Beyond these chapters, the book unfolds various benevolent use cases, providing insights for non-profit organizations contemplating blockchain implementations for their operations related to underserved and vulnerable demographics.

NAVIGATING THE AUDIENCE LANDSCAPE

The relevance of this book extends to a diverse audience, including the following:

IT and Business Professionals, especially those familiar with blockchain concepts seeking a deeper understanding of data architecture, privacy, security and operational risks related to blockchain platforms.

Undergraduate and Graduate Business Students, especially those specializing in IT, accounting and information systems auditing aiming to grasp the potentials and caveats of blockchain technology.

Business Leaders, particularly those leading benevolent, non-profit organizations interested in exploring blockchain possibilities for their own operations.

Accountants, Information Systems and Internal Audit Professionals looking to deepen their comprehension of the attributes and limitations of this emerging technology.

Individuals Aspiring for a Blockchain Career Path seeking to break into the blockchain technology ecosystem and understanding its foundational aspects.

University Libraries aiming to enrich their blockchain book collection with contemporary and insightful references.

Undergraduate or Graduate IT Faculty looking to supplement their course modules with a relevant, management-focused comprehensive blockchain technology textbook.

HOW TO DERIVE MAXIMUM KNOWLEDGE FROM THIS BOOK

To derive maximum value from this book, it is advisable to have a basic understanding of blockchain technology. The preface chapter's section, *Expanding the Exploration of Blockchain Technology*, can act as a quick preliminary assessment. If the content in this section resonates, readers can seamlessly progress through this book. For those not acquainted with blockchain fundamentals, *The Auditor's Guide to Blockchain Technology* will offer a starting point to derive maximum benefit from this book.

Both *Leveraging Blockchain Technology* and *The Auditor's Guide to Blockchain Technology* are also designed as reference books, enriching the library of IT professionals. The titles nicely complement each other, providing for a rather comprehensive blockchain education.

HOW IS THIS BOOK ORGANIZED?

Each chapter follows a simple, yet structured format to help ensure a coherent and insightful reader exploration, with its following main sections:

The *Abstract* provides a high-level summary offering readers a quick insight into the subtopic and findings. This aids in selecting chapters based on individual interests.

The *Background* section helps set the stage for deeper exploration as this section provides the necessary context, thereby laying the foundation for the ensuing discussion.

The *Discussion* presents research-based information and insights into various facets of blockchain technology.

The *Conclusions and Recommendations* section summarizes key concepts discussed in each chapter. This section also offers related good practices and recommendations derived from the findings.

The *References* provide researchers with a comprehensive list of relevant readings and reference materials for further exploration and use in their own research projects.

In the chapters that follow, let us immerse ourselves in the various dimensions of blockchain technology by exploring blockchain governance, operational risks, privacy implications, security measures, financial reporting considerations, as well as a number of benevolent use cases. This book is not merely a compilation of facts but a journey into the intricate and evolving landscape of blockchain technology. As we traverse this exploration, let us unravel the layers of blockchain's potential and its far-reaching impact on the digital era.

I wish you all the very best in your blockchain exploration journey…

Shaun Aghili
Management Professor
Information Systems Security and Assurance Management Department
Concordia University of Edmonton
Summer 2024

Editor

Shaun Aghili, an award-winning management professor, is the lead faculty of Concordia University of Edmonton's Master of Information Systems Assurance Management (MISAM). MISAM is Canada's only graduate-level program in information systems auditing.

With numerous professional designations to his name, Dr. Aghili is a certified management accountant (CMA), an internal auditor and risk assurance specialist (CIA, CRMA), a fraud examiner (CFE), an information systems auditor (CISA), and an information systems and cloud security professional (CISSP-ISSMP, CCSP). Shaun has also completed several blockchain certifications, such as Certified Blockchain Expert and Certified Blockchain Security Professional.

Dr. Aghili has authored over 90 published articles, book chapters and conference proceedings, including his latest books titled *Fraud Auditing Using CAATT: A Manual for Auditors and Forensic Accountants to Detect Organizational Fraud* (2019) *and The Auditor's Guide to Blockchain Technology: Architecture, Use Cases, Security, and Assurance (2022).*

Contributors

Lilian Behzadi
Master of Information Systems Security
  Management Graduate
Concordia University of Edmonton
Alberta, Canada

Aditya Bhatt
Master of Information Systems Security
  Management Graduate
Concordia University of Edmonton
Alberta, Canada

Alamgir Faruque
Master of Information Systems Security
  Management Graduate
Concordia University of Edmonton
Alberta, Canada

James Joseph
Master of Information Systems Security
  Management Graduate
Concordia University of Edmonton
Alberta, Canada

Priyank Mehta
Master of Information Systems Security
  Management Graduate
Concordia University of Edmonton
Alberta, Canada

Ebubechi Okpaegbe
Master of Information Systems Security
  Management Graduate
Concordia University of Edmonton
Alberta, Canada

Adebimpe Onamade
Master of Information Systems Security
  Management Graduate
Concordia University of Edmonton
Alberta, Canada

Ami Padia
Master of Information Systems Security
  Management Graduate
Concordia University of Edmonton
Alberta, Canada

Pramitha Pinto
Master of Information Systems Security
  Management Graduate
Concordia University of Edmonton
Alberta, Canada

Kolawole Salako
Master of Information Systems Security
  Management Graduate
Concordia University of Edmonton
Alberta, Canada

Kesha Sisodia
Master of Information Systems Security
    Management Graduate
Concordia University of Edmonton
Alberta, Canada

Tahziba Tabassum
Master of Information Systems Security
    Management Graduate
Concordia University of Edmonton
Alberta, Canada

Ankita Vashisth
Master of Information Systems Security
    Management Graduate
Concordia University of Edmonton
Alberta, Canada

Blockchain data structure, governance, risk, privacy, security and digital assets considerations

Chapter 1

Blockchain data and architecture considerations

Ami Padia and Priyank Mehta

BACKGROUND

In the contemporary landscape, most transactions are presently overseen by intermediaries, primarily banks. These financial entities often impose substantial fees for their services, particularly in terms of transfer fees. A potential remedy for this reliance on third-party intermediaries is presented by blockchain technology. This innovative approach entails a distributed, decentralized digital ledger designed to record transactions and securely store data. Originally developed for Bitcoin cryptocurrency, blockchain technology has progressively found applications in diverse domains, including healthcare, supply chain management, humanitarian aid and more. Transactions are recorded in blocks, and a new block is appended to the existing chain upon the completion of a new transaction. This technology ensures transactional security, persistence, decentralization, anonymity and data integrity [1].

A blockchain functions as a distributed digital ledger spread across a network. Each block within the chain encapsulates a set of verified and encrypted transactions. Once incorporated into the chain, it remains unalterable. The security features of blockchain technology are underpinned by asymmetric cryptography, which involves the use of two distinct keys for data encryption and decryption along with distributed consensus algorithms employed in distributed systems to reach agreement among nodes or participants. The implementation of blockchain extends beyond transactions, encompassing areas such as smart contracts, Internet of Things (IoT), security services and public services. The immutability of blockchain ensures that once data is entered into blocks, it remains unaltered. This technology fosters a high level of security and transparency, granting all network participants access to identical information and the ability to verify transactions without relying on a central authority. The initial successful application of blockchain technology, exemplified by Bitcoin, established a decentralized environment for cryptocurrency. However, while blockchain technology proves suitable for cryptocurrency transactions, it encounters inherent limitations and challenges that necessitate further research [1, 2].

DISCUSSION

As mentioned, blockchains are a type of distributed ledger technology (DLT) that facilitates secure and transparent peer-to-peer transactions without reliance on a central authority. Transactions are systematically recorded in blocks, forming an immutable and chronological chain. This versatile technology finds applications across various industries, including finance, supply chain management and healthcare. The main types of blockchains are outlined below (Table 1.1).

Table 1.1 Blockchain architectural designs

Blockchain type	Features or attributes	Limitations
Public	Anyone can join the network, and it is highly decentralized. Provides high security and privacy [3].	Requires a lot of energy as it often uses proof-of-work. Scalability issues on a lower transaction per second [3].
Private	Faster and higher scalability. Transaction per second (TPS) rate higher for a smaller number of nodes [3].	Has weak security and is easy to hack entire private blockchain [3].
Hybrid	Because of closed ecosystem, resistant to 51% attacks on a high network performance, making it more scalable and cheaper [5].	Has a closed ecosystem. Does not provide complete transparency [5].
Consortium	Provides higher transaction speed and scalability and gives control to organizations [5].	Lack of transparency and increased risk of attack when a few nodes are compromised [5].
Permissioned	Typically applies to private blockchains. Requires access control or permissions to read and/or write. Handles fewer transactions compared to permissionless blockchains, offering higher performance and scalability [6].	Less transparent as controlled by organizations and has fewer participants. Less participants may lead to higher chances of collusion and corruption of data [6].
Permissionless	Highly transparent. Provides more security resilience. Fully decentralization allows more participants [6].	Less user privacy as no required permission to join. Less energy efficient as it uses a lot of resources on a network [6].

Private blockchains

Private blockchains operate within a restricted network where only an authorized group of individuals can contribute data. This controlled network is overseen by a single identity, making it suitable for small businesses, organizations and enterprises. Access to the private blockchain is limited to permitted members, ensuring secure transactions within departments and branches. The system administrator manages permissions, granting access to users who seek to join the network. Private blockchains function in closed systems, providing security, authorizations, permissions and controlled accessibility. Key features of private blockchains include privacy, high efficiency, faster transactions, better scalability and improved speed. Experts suggest that private blockchains are specifically designed for applications such as digital identity management, voting systems, supply chain management and asset ownership. Unlike public blockchains, private blockchains do not require cryptocurrency as a fundamental element. Transactions related to nodes can only be added or accessed by authorized users within the restricted network [3].

Public blockchains

Public blockchains stand out as a major type due to their operations on open and decentralized networks. In this system, all transactions are recorded on a public ledger that is transparent and accessible to anyone. Unlike private blockchains, public blockchains are non-restrictive and allow for a distributed ledger where users can access information without requiring any permission. The consensus mechanisms commonly employed in public blockchains include proof-of-work and proof-of-stake, ensuring validation and agreement on transactions by participants.

In a public blockchain, historical and contemporary transaction records are openly available for users to verify. This type of blockchain is instrumental in mining operations and the exchange of cryptocurrencies. Notable examples of public blockchains include *Bitcoin* and *Ethereum*, which are widely used for various transactions. A key advantage of public blockchains lies in their decentralized nature, making it challenging for any single entity or group to control the network. This inherent decentralization enhances security and resistance to censorship and other forms of attacks.

While public blockchains have traditionally been associated with cryptocurrency transactions, their applications are expanding. Smart contracts, decentralized applications (dApps) and supply chain management are emerging uses of public blockchains. The distinctive features of public blockchains include high security and privacy, an open and flexible environment, anonymous transactions, absence of strict regulations and complete transparency [2, 3].

Hybrid blockchains

A hybrid blockchain represents a fusion of the advantages associated with both public and private blockchains. This type of blockchain incorporates public and private elements, combining the transparency and decentralization of public blockchains with the controlled security features of private blockchains.

In the public aspect of a hybrid blockchain, transactions are recorded on a decentralized ledger, mirroring the characteristics of a traditional public blockchain. This transparent and immutable ledger allows anyone to participate in the network and view transactions. Conversely, the private aspect of the hybrid blockchain introduces a more controlled and secure environment. Access to the private blockchain is restricted to a limited group of participants, carefully selected and granted permission to join the network. This restricted access enhances control over participant inclusion and the transactions recorded on the blockchain [3, 4].

The architecture of a hybrid blockchain permits seamless integration with public blockchains. Transactions are often initially validated within the network, and users can subsequently publish them on the open blockchain for verification. This dual-layered approach, by incorporating enhanced hashing and employing additional nodes for confirmation, serves to elevate both the security and transparency of the overall blockchain network [5].

Consortium blockchains

Consortium blockchains represent a semi-decentralized form of blockchain technology jointly owned and operated by a group of organizations or entities. In contrast to public blockchains, which allow open participation, consortium blockchains are permissioned, requiring approval from the consortium for individuals or organizations to join the network.

In a consortium blockchain, participating organizations collaboratively share the responsibility of network maintenance and transaction validation. This structure offers greater control and privacy compared to public blockchains while still retaining the advantages of decentralization and transparency. Consortium blockchains find practical utility in industries where collaboration and information sharing among multiple organizations are crucial, such as in supply chain management, finance and healthcare.

By leveraging a consortium blockchain, these organizations can securely and transparently share data without relying on intermediaries or third-party validators. Consortium blockchains strike a balance between the decentralized and transparent nature of public blockchains and the controlled and private characteristics of private blockchains. This makes

Table 1.2 Permissioned versus permissionless blockchains

Permissionless blockchain	Permissioned blockchain
Open network configuration. Anyone can join the network.	Closed network configuration. Only permitted members can join and interact.
Also known as public.	Also known as private.
Gives the transparency of transactions.	Controls the transparency of transactions based on organizations.
Can be developed with open-source tools.	Can be developed by private entities.
Mostly anonymous.	Not anonymous.
Privacy is dependent on technological limitations.	Privacy is dependent on governance decisions.
Less secure because permission is not needed.	More secure because access permission is required.
Support peer-to-peer architecture.	Support business-to-business architecture.
May be used in digital trading, donation and crowdfunding.	May be used in supply chain tracking systems, verification of identity and claim settlements.

them a preferred choice for organizations seeking to harness the capabilities of blockchain technology while addressing specific privacy and control requirements [3, 5].

Ripple, a prominent cryptocurrency, serves as an illustrative example of a permissioned blockchain. Unlike Bitcoin and Ethereum, which operate as permissionless networks, Ripple employs a permission-based framework for its network participants.

The drawbacks associated with permissionless blockchains, such as Bitcoin and Ethereum, predominantly center around performance issues. A critical challenge for permissionless blockchains is the substantial energy and computing power required for the consensus process. The resource-intensive nature of consensus mechanisms poses a significant obstacle.

It is essential to clarify that the distinction between "public" and "private" blockchains does not align with the traditional understanding of these terms. In the context of blockchains, a permissionless blockchain is considered public, and a permissioned blockchain is considered private. However, the definition of "public" in this context refers to being open for use rather than open for reading the transaction history. It is important to note that a "private blockchain" can be "public to view" without being "public to use." This distinction highlights the nuanced nature of blockchain accessibility and usage (Table 1.2) [7, 8].

Layers of blockchain technology

In the realm of blockchain technology, information is securely stored on a distributed ledger. This ledger, governed by a predetermined protocol, relies on a consensus mechanism involving multiple nodes across the network to validate transactional data. The layered design of blockchain technology facilitates transaction authentication [9].

Layer 0 constitutes the foundational physical layer supporting the blockchain network. Comprising hardware infrastructure such as servers, network devices and data centers, this layer establishes the groundwork for the blockchain's operation. Serving as the network architecture underlying the blockchain, Layer 0 is pivotal to its functionality. This physical layer provides essential computing power and network connectivity enabling network nodes to communicate and validate transactions. In proof-of-work blockchain networks like Bitcoin, Layer 0 supports mining nodes responsible for the computational work necessary to add new blocks. Moreover, Layer 0 facilitates inter-chain operability, allowing communication between different blockchains. It plays a crucial role in resolving scalability issues, often utilizing native tokens for participation and development. Cryptocurrencies like *Polkadot* and *Avalanche* exemplify Layer 0 implementations [9].

Layer 1 constitutes the foundational protocol layer that governs the entire operation of the blockchain network. The majority of essential operations that sustain the core functions of a blockchain network – such as dispute resolution, consensus mechanisms, programming languages, protocols and limitations – are executed at Layer 1.

This layer encompasses the core blockchain protocol, delineating the rules for validating transactions, incorporating new blocks into the blockchain and achieving consensus among network nodes. The Layer 1 protocol is strategically designed to establish a secure and decentralized network that operates autonomously without relying on a central authority or intermediary. A key aspect of the Layer 1 protocol is its consensus mechanism defining how network nodes collectively agree on the state of the blockchain.

Scalability challenges often arise due to the substantial workload managed by Layer 1. The computational power required to solve and add blocks to a blockchain increases with the growing number of users resulting in higher fees and prolonged processing times. However, emerging solutions like proof-of-stake (PoS) and the introduction of sharding (division of computing operations into smaller parts) provide partial mitigation for scalability issues. Noteworthy cryptocurrencies operating at Layer 1 include *Ethereum*, *Bitcoin* and *Solana* [9, 10].

Layer 2 plays a crucial role in enhancing the efficiency of blockchain productivity by demanding increased processing power. However, this improvement comes with the need for additional nodes, which can potentially congest the network. The addition of nodes is essential to preserve the decentralized nature of a blockchain, but adjustments made to scalability, decentralization, or throughput in Layer 2 can impact other Layer 1 factors. Consequently, expanding Layer 1 without transferring all processing to Layer 2 added on top of Layer 1 becomes impractical.

The integration of third-party solutions with Layer 1 is made feasible by enabling Layer 2. This layer acting as a second network updates Layer 1 and assumes control over all transactional validations. Within the blockchain ecosystem, Layer 2 is positioned above Layer 1 and maintains frequent communication with it. In contrast, Layer 1 is responsible solely for overseeing the creation and addition of new blocks to a blockchain.

Layer 2 solutions contribute significantly to accelerating the speed and efficiency of blockchain transactions. They achieve this by allowing multiple transactions to be executed off-chain before settling on the Layer 1 blockchain. However, the introduction of Layer 2 also introduces new challenges and complexities particularly regarding the security and trustworthiness of off-chain transactions [9].

Layer 3 represents the final tier in the blockchain ecosystem, where participants ultimately engage with user interfaces (UI). The primary focus of Layer 3 is to offer simplicity and user-friendly interactions with both L1 and L2.

This layer serves a dual purpose by providing intra- and inter-chain operability facilitating decentralized exchanges, liquidity provisioning, staking applications and user interfaces. Particularly, decentralized applications (dApps) serve as a form of Layer 3 interface translating blockchain technology into real-world applications. These applications bring about tangible utility and demonstrate the transformative potential of blockchain technology in various industries.

Layer 3 solutions harness the inherent attributes of the blockchain – security, transparency and immutability – to develop innovative applications that address longstanding issues. As a result, Layer 3 is where the real-world impact of blockchain technology is most profoundly felt, introducing novel solutions across diverse industries. However, the implementation of Layer 3 solutions introduces new challenges including regulatory and legal considerations, as well as the necessity for meticulous attention to user privacy and security [9].

Layers of blockchain technology – expanded perspective

Alternatively, according to some blockchain experts, the blockchain technology ecosystem is composed of five distinctive layers, each playing a crucial role in its overall functionality.

The Infrastructure or Hardware Layer constitutes the foundational technological components that facilitate the operation of the blockchain system. Essential to this layer is the secure storage of blockchain data on a dedicated server. When users attempt to access blockchain applications or navigate the Internet, computers initiate requests to access this stored data from the server. Peer-to-peer (P2P) networks form the backbone of blockchain platforms, enabling seamless and rapid connections between nodes for information sharing. This exchange of data is facilitated through the client–server architecture, envisioning a vast network of interconnected devices interacting and exchanging information. Notably, the utilization of distributed ledger technology enables each node to monitor transactions and engage with other nodes in the network [9, 10].

The Data Layer within blockchain technology pertains to the arrangement and organization of the information stored on the blockchain. This layer assumes responsibility for upholding the integrity and security of the data, ensuring efficient accessibility and facilitating verification by network participants.

At the core of the data layer is the concept of the *Genesis Block*, which serves as the inaugural block in the blockchain. Subsequent blocks are linked to the Genesis block through an iterative mechanism perpetuating the growth of the blockchain. Each transaction is digitally signed using the sender's wallet private key, a key visible only to the sender. This digital signature encrypted for heightened security safeguards the data from unauthorized access or alterations. This cryptographic process, known as *finality* in the realm of blockchain technology, virtually eliminates the possibility of tampering with data access.

The Network Layer plays a pivotal role in the blockchain ecosystem by facilitating the exchange of transaction data among numerous nodes within the peer-to-peer (P2P) infrastructure. This layer – also referred to as the *Propagation Layer* – is instrumental in ensuring fast and effective inter-node communication. The efficient functioning of the blockchain network relies on each node's ability to locate and connect with other nodes swiftly. The network layer oversees critical processes such as block creation, block addition and node detection, with nodes being responsible for processing transactions on the blockchain [9].

The Consensus Layer stands as an indispensable component for the functionality of blockchain platforms regardless of the specific blockchain technology being utilized (e.g., Ethereum, Hyperledger). This layer assumes a central role in ordering, validating and ensuring the correct sequence of blocks in the blockchain [10]. Transaction authentication, one of the most crucial functionalities in blockchain, is carried out by the consensus layer. Without this layer, the system would fail to operate effectively as transaction validations would be lacking.

The consensus layer executes the protocol that mandates a specific number of nodes to verify a single transaction. This decentralized approach involves multiple nodes processing each transaction independently, all converging on a unanimous conclusion to accept it as legitimate. This distributed nature, known as a consensus mechanism, preserves the decentralized character of the blockchain. While the simultaneous creation of multiple blocks by various nodes may lead to a temporary branching in the blockchain, the consensus layer ensures the resolution of such conflicts, ensuring that a single chain block update prevails [9].

Moreover, the digital signature not only secures the data but also protects the identity of the sender or owner. The signature is legally binding to the signer enhancing the overall security of the transaction and ensuring that the identity associated with the signature is irrefutable [10].

The Application and Presentation Layer within the blockchain architecture encompasses a suite of critical components, including smart contracts, chaincode and decentralized applications (DApps). Further subdivision occurs at the level of application layer protocols.

At the forefront of user interaction with the blockchain network; and, as mentioned, the application layer comprises smart contracts, chaincode and DApps. This layer extends into various application layer protocols. It is the software layer through which end-users engage with the blockchain network featuring user interfaces, frameworks, application programming interfaces (APIs) and scripts. The backend technology supporting these applications is rooted in the blockchain network, with interactions facilitated through APIs. Notably, the applications direct the execution layer.

The presentation layer, on the other hand, is concerned with the user interface shaping the way end-users interact with the blockchain network. It involves the visual and graphical aspects of blockchain applications, enhancing the user experience and accessibility.

The Execution Layer, distinct from the application layer, comprises chaincode, smart contracts and underlying logic. While a transaction traverses from the application layer to the execution layer, approval and execution occur at the semantic layer. Applications guide the execution layer by issuing commands and upholding the determinism of the blockchain. This layer plays a pivotal role in translating high-level transactions initiated by users into executable code within the blockchain network (Table 1.3) [10].

Blockchain data type and data structure

Data Type is defined by the rules governing the classification of the type of value a variable may contain. It also dictates the mathematical, logical or relational operations that can be performed on that value without encountering errors [11]. Ongoing research in the blockchain domain aims to develop a data type that minimizes memory usage while executing precise computations surpassing the capabilities of traditional standard data types.

The Transaction Data Model outlines how transactions are organized and recorded in the network. Below are the typical components of a transaction data model in a blockchain:

Input refers to the source of the transaction, such as the wallet or address sending the funds.

Output represents the destination of the transaction, such as the wallet or address receiving the funds.

Amount signifies the value of the transaction, usually denominated in the blockchain's native currency.

Table 1.3 Mapping of the two proposed BC layers

Level	Level	Functionality
0	Infrastructure or Hardware Layer	Provides the foundational elements for the blockchain
I	Data Layer	Maintains the programming, consensus mechanism and data
2	Network Layer	Provides better scaling and integration with third-party tools
3	Consensus Layer	Hosts dApps and other user-facing applications
N/A	Application Layer	Enables user-facing applications at the front end

The Transaction ID is a unique identifier assigned to the transaction by the network facilitating tracking and verification.

The Timestamp reveals the date and time when the transaction was initiated.

Fees refer to transaction costs paid by the sender to incentivize network validators for including the transaction in the blockchain.

A *digital signature* is created by the sender as proof of authorization for the transaction.

Understanding these components provides insight into how transactions are structured and ensures the integrity and security of the blockchain network [12].

Extended components in the transaction data model

Beyond the core components outlined earlier, certain blockchain networks may incorporate additional fields, such as metadata or smart contract code. These supplementary elements offer enhanced context or functionality to transactions. The comprehensive transaction data model is indispensable to blockchain operations providing a standardized format for recording and verifying transactions. This ensures the overall integrity and security of the network [12].

An alternative transaction model, known as the *Account-based Model*, is notably employed by Ethereum. In this model, assets are represented as balances within accounts akin to traditional bank accounts. The account-based model operates with two distinct types of accounts:

Private Key-Controlled User Account, also referred to as Externally Owned Account (EOA), is under the direct control of the user via a private key. Each account address is derived from a public key, shareable within the blockchain network. Access to these accounts is secured by a private key, often considered a PIN, which must be kept confidential and known exclusively to the account owner [13].

On the other hand, a *Contract Code-Controlled User Account* relies on smart contract code to manage its operations. A smart contract is essentially a specific set of code designed to execute when predetermined conditions are met [14]. Transactions cannot be initiated directly from these accounts. Instead, when a transaction is sent to a contract code-controlled account, the account triggers various functions within other contracts.

Each account within the account-based model encompasses the following fields:

Balance represents the amount of balance held in the account.

A *Nonce* is a term used to refer to a number that is used only once. It is a cryptographic value that is included in the data of a block during the mining process. The primary purpose of a nonce is to adjust the block's hash in such a way that the resulting hash meets certain criteria, typically a specific number of leading zeros.

Storage pertains exclusively to Contract Code-Controlled User Accounts. This field serves as a permanent data store for the respective account.

Code comprises the set of instructions executed by a contract code-based account upon receiving a transaction. This field remains empty for EOAs [13].

The data structure of the blockchain is designed as a chain structure designed to detect tampering or data modification. While different blockchain platforms may employ varying data structures, the fundamental structure remains consistent. Taking Bitcoin as an example, it features a block header and block body with each block identifiable by its hash. Hash blocks are generated using the SHA256 cryptographic algorithm on the block header, also known as the *parent block*. The header block stores the pre-block hash, nonce and Merkle root. Common hash structures include the Merkle tree and block list [12, 15, 16].

Merkle tree

The Merkle Tree, proposed by Ralph Merkle and employed by Bitcoin, serves as a crucial data structure within the blockchain architecture. In this tree structure, each node represents a hash value, specifically the SHA256 hash value of a transaction. The process involves combining the values of two leaf nodes and then computing parent node values through hashing. This hashing operation continues until the generation of the Merkle root. The Merkle root acts as a means to detect data tampering, ensuring the integrity of the data.

Utilizing the Merkle root, it becomes possible to verify transactions based on simplified payment verification (SPV), solely focusing on the path from the transaction node to the Merkle root without involving other nodes in the tree. When there are N transactions in a block, a maximum of 2log2(N) hash operations can be performed to verify the existence of the transaction. This form of verification, which does not require full data participation, is advantageous for applications like e-wallets.

For instance, in Ethereum, the Merkle root is calculated using the Merkle Patricia tree. This approach is suitable when there is a change in data status and the transaction data in the block remains relatively unchanged. When constructing a new block, only the statuses of the accounts that have changed need to be calculated while those with no change in status can be added as is (Figure 1.1) [12].

The block list

In the block list method, the hash value of the block header is generated by applying the SHA256 hash on metadata. This metadata encompasses essential information like the previous block hash, Nonce and Merkle root within the block header. The blocks are interconnected forming a linked structure known as a block list. With the previous block hashes stored in the header, it becomes feasible to verify the integrity of all blocks by performing block hashing.

By utilizing this block hashing technique, data verification becomes straightforward, allowing for the detection of modifications in one or more blocks downloaded from an untrusted node. This method ensures the tamper detection of data by validating all previous blocks and creation blocks against their respective hash values.

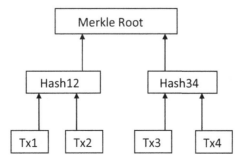

Figure 1.1 Merkle root [12].

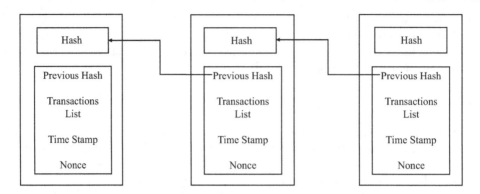

Figure 1.2 Block list [12].

Moreover, the transaction Merkle root present in the block header facilitates the verification of potential tampering with both the transaction data and block header. As this method relies on previous block hashes, any alteration in one block triggers changes in all other blocks, creating a chain effect in hash pointers. This ensures timely detection of data tampering offering a robust mechanism for maintaining the integrity of the blockchain (Figure 1.2) [12].

Data policy

Public blockchains, operating on decentralized networks in contrast to private centralized blockchains, lack a central authority for data policy. Instead, the blockchain policy is determined by the consensus mechanism employed within the network, and rules are encoded through smart contracts. Despite the absence of a central governing body, there exist fundamental principles guiding and regulating blockchain applications.

Tech neutrality is a crucial aspect of blockchain policy emphasizing that policies and regulations should neither favor nor discriminate against any specific technology. Policymakers must embrace technologies in a manner that ensures a tech-neutral environment for diverse applications. While certain projects may find greater benefits in utilizing distributed and tamper-resistant blockchains, others might operate more efficiently through traditional centralized methods. For example, a blockchain might be indispensable in a digitization project requiring inputs from various sources ensuring independence from the authority of any specific company. To maintain fairness, different technologies supplying identical goods and services should be subject to the same regulatory standards. It is important to note that not all blockchain applications warrant regulation, as their unique characteristics may render traditional regulatory frameworks unnecessary.

To level the playing field between established and emerging technologies or business models, regulators should scrutinize existing regulations and assess their applicability to evolving applications. Recognizing the diversity of technological applications is paramount as unregulated digital currencies may not pose the same concerns about money laundering as traditional banking reliant on centralized databases. In instances of technological divergence, regulators should design nuanced regulations that consider the unique threats or irrelevance to specific applications.

Unfortunately, regulators have not consistently maintained a tech-neutral approach when dealing with blockchain-based applications. For instance, organizations involved in virtual currencies often operate similarly to those engaged in mobile payments and international transfer services [16].

Promoting blockchain adoption and deployment

Governments at national, subnational and local levels play a pivotal role in fostering the widespread adoption of blockchain technology. Acting as early adopters themselves, governments can encourage others to follow suit and invest in blockchain, thereby mitigating associated risks. Initiatives should encompass leveraging blockchain solutions from companies to enhance various operational aspects, including reporting, transactions, asset tracking, supply chain management, procurement and budgetary decision-making.

As an illustration, the U.S. Department of Health and Human Services has successfully implemented a blockchain application to streamline service procurement and cloud modernization resulting in cost savings. To support such initiatives, governments may need to revise existing procurement procedures. Currently, the regulations and procedures governing how U.S. federal agencies adopt technologies are deemed outdated and burdensome. For instance, federal agencies can acquire goods and services through various channels, such as open-market acquisitions and contracts like the General Services Administration (GSA) Multiple Award Schedule contracts.

However, the procurement process for blockchain solutions is constrained as federal procurers are restricted from acquiring such solutions unless they are established by government IT suppliers on an existing blanket purchase agreement (BPA). This limitation, executed through open-market purchases, often proves time-consuming and is subject to extensive scrutiny, leading to delays of six months to a year. This lag in procurement is particularly challenging for rapidly evolving technologies like blockchain.

A suggested improvement involves the GSA creating a dedicated section to assist agencies in procuring blockchain technology and streamlining procedures. Establishing a core team of government procurement experts for blockchain within GSA can further enhance efficiency by advising and facilitating blockchain-related procurement at other agencies. This proactive approach ensures that the federal government utilizes blockchain, effectively preventing wasted resources on projects ill-suited for the technology.

In parallel, for regulatory compliance, businesses are increasingly adopting technologies like blockchain termed as *Regtech*. These solutions not only aid firms in complying with regulations but also elevate the quality and efficacy of oversight by providing regulators with access to modern reporting and analytics infrastructure. It is imperative for authorities to collaborate with businesses ensuring that these processes do not favor specific companies or technological advancements or harm consumers or pose systemic risks [16].

Supporting blockchain research and development

Government investment has historically played a pivotal role in advancing technological frontiers, as exemplified by the development of the Internet and smartphones. In the early stages of technology research, emphasis is often placed on conceptualization rooted in theories rather than tangible products. To enhance the efficiency of blockchain technology, governments should allocate funds or invest in the Research & Development (R&D) sector specifically dedicated to blockchain applications. This focus allows R&D efforts to address challenges associated with blockchain technology, including the identification of security threats, development of more efficient consensus mechanisms, enhancements in cryptography, scalability and other pertinent aspects. Additionally, R&D can extend its scope to improve associated technologies such as quantum computing, potentially amplifying the capabilities of blockchain applications. However, ensuring enforcement of issues like managing intellectual property rights over public blockchains will necessitate extensive study and collaboration between the public and commercial sectors [14, 16].

Developing flexible regulatory frameworks to encourage experimentation

The U.S. regulatory environment, particularly in mature sectors like financial regulation, often impedes the development of innovative blockchain-based products and services due to a lack of clear categorization within existing legal frameworks. This is particularly evident in the case of decentralized applications (dApps). The uncertainty surrounding potential enforcement actions discourages blockchain-based firms from launching in the United States.

To address this issue, regulators need to establish administrative procedures that help understand and define blockchain-based goods and services. Cryptocurrency assets exhibit variations in technology, transaction speed, utility, decentralization and other characteristics. Regulators should establish specialized committees and working groups to thoroughly investigate and categorize these new goods and services, providing a clearer understanding of cryptoassets.

Instead of relying solely on enforcement actions, regulatory authorities can employ alternative administrative tools to provide relief to corporations. One such approach is issuing no-action letters, where companies submit applications along with data-sharing agreements for a specific product or service. Upon approval, the agency commits not to take legal action against the product or service. This approach allows regulators to communicate acceptable behaviors to the market through no-action letters, fostering the development of financial services in response to regulatory signals.

Additionally, regulators can establish regulatory sandboxes to facilitate the testing of products or services in a controlled environment under regulatory supervision. This offers a safe harbor with no-enforcement-action relief, enabling companies to collaborate with regulators in the pursuit of more innovative solutions. While self-regulatory frameworks complement these efforts, they are most effective when supported by regulatory enforcement and characterized by minimal degrees of control [16].

Data ownership considerations in blockchain technology

A database functions much like an address book, systematically organizing and storing data for quick access. Blockchains allow the recording of data exchanged over the Internet along with its sources and recipients, creating a precise, enduring and cost-effective database. The pioneering use of blockchain technology came with the introduction of Bitcoin, a virtual currency functioning similarly to physical currency based on bank accounts.

In traditional banking, accounts meticulously record credits and debits, maintaining accuracy and integrity. Blockchain technology applies a comparable concept to virtual currency accounts, known as *wallets*. Unlike traditional banking, blockchain operates without a central authority relying on algorithms to maintain an indisputably reliable record. This trustworthiness, akin to physical currency, establishes each piece of data's ownership to an account holder.

The innovation of blockchain technology extends to individuals more control over their personal data through the concept of data ownership. Users can decide what data to share, and with whom, and establish permissions and access controls, enhancing privacy and security. Additionally, blockchain enables individuals to monetize their data by selling or licensing it to third parties, creating new revenue streams.

However, amid the benefits of data ownership in blockchain technology, challenges and drawbacks exist. Managing one's own data involves overseeing permissions, access controls, accuracy and protection against unauthorized access. For those unfamiliar with blockchain or data management procedures, this can pose difficulties. Currently, the legal frameworks supporting data ownership in blockchain are lacking, presenting challenges in establishing ownership rights, upholding agreements and safeguarding against theft or fraud. This absence of legal foundations introduces uncertainty and risk for individuals managing their data on the blockchain.

Furthermore, the decentralized nature of blockchain technology means there is no central organization responsible for managing or recovering lost data. If an individual misplaces their private key or password, they may permanently lose access to their data. This poses a significant risk for those relying on blockchain technology to store crucial information [17] [18].

Data creation: mining for cryptocurrencies

Cryptocurrency mining is a process that is both time-consuming and occasionally profitable. It involves investors contributing computational power to validate and add transactions to the blockchain. In return for their efforts, miners receive cryptocurrency tokens, making mining an attractive avenue for investors. Bitcoin miners, for instance, are rewarded with Bitcoin for completing "blocks" of verified transactions added to the blockchain.

The mining process requires specialized hardware, such as a GPU (graphics processing unit) or an ASIC (application-specific integrated circuit). While early Bitcoin mining could be done with standard home computers, the increasing difficulty of mining, adjusted over time, has made this impractical. The competition among mining rigs influences the speed at which the hashing puzzle is solved. To maintain the blockchain's efficiency, the Bitcoin network aims to produce a block approximately every ten minutes, adjusting difficulty to balance the mining power engaged.

Data archiving in blockchain

Data archiving involves moving inactive data to secondary storage, preserving its importance for future reference or regulatory compliance. In blockchain, data archiving prevents unwanted data accumulation in blocks, necessitating the transfer of non-useful data to secondary blocks. Utilizing blockchain for data archiving offers several advantages, notably the creation of a decentralized storage system.

Decentralized storage in blockchain ensures data is distributed across a network of nodes, reducing vulnerability to hacking or data loss. This decentralized nature eliminates a single point of failure, enhancing resilience against attacks. Blockchain technology also provides robust data security and privacy. Data can be encrypted and securely stored on the blockchain, restricting access to authorized individuals. The absence of third-party intermediaries further diminishes the risk of data breaches or security issues in the archiving process [19].

Data retention in blockchain

Data retention involves storing data for a specific period. In the context of blockchain, it determines how long a block can retain information. Comparable to a parking system with time limits, blockchain requires data to move to the next block when the time limit for one block expires. Each block in the chain can store less than 1 MB of data, and any excess data beyond this limit is transferred to the next block. This ensures that the blockchain remains efficient and does not store excessive information indefinitely.

As a decentralized network, blockchain allows authorized users to access information globally at any time. While this facilitates information sharing and access, it's crucial to note that blockchain technology is not a one-size-fits-all solution for data storage. Without proper data retention practices, blockchain may accumulate excessive information, leading to operational inefficiencies, increased costs and potential legal and security risks [20].

Data storage in blockchain

Blockchain data storage employs a distributed database, decentralizing data storage across a network of computers. Each node within the network continually updates and synchronizes the database with other nodes. Data in a blockchain is organized into blocks, each containing a collection of transactions or other data and linked chronologically by referencing the preceding block in the chain.

Decentralized blockchain data storage eliminates single points of failure and vulnerabilities, ensuring high security and resistance to data loss or hacking. It is open to all authorized users and transparent. Beyond cryptocurrency transactions, blockchain data storage has diverse applications, including supply chain management, voting processes and identity verification [21].

In the process of blockchain storage, files undergo partitioning, known as sharding, to prevent data loss in case of transmission errors. Each shard is replicated and encrypted with a private key to restrict unauthorized access and then distributed globally among decentralized nodes. Interactions are recorded in the blockchain ledger, ensuring confirmation and synchronization of transactions across nodes. Blockchain storage intends to perpetually record interactions. Once recorded, the data is immutable, meaning it cannot be altered [20].

Data storage in cloud versus blockchain

In a cloud storage system, users have the ability to modify previously saved data, allowing for changes that can impact how others perceive the information. In contrast, blockchain enforces immutability, preventing users from altering existing data. Entries in a blockchain cannot be deleted or changed, serving as a robust defense against data tampering.

Cloud storage systems offer the flexibility of both public and private data, providing varying levels of accessibility. However, data stored on a blockchain is inherently public. Blockchain storage also allows users to customize settings such as retrieval speed and redundancy, potentially resulting in faster and more tailored storage systems.

Data sharing in blockchain

Data sharing within the context of blockchain technology involves enabling multiple parties to access and share data securely and transparently. Blockchain employs a decentralized and immutable ledger system, ensuring the integrity of shared data and transactions. In a blockchain network, data is organized into blocks connected through cryptography, forming an unalterable chain of blocks.

The decentralized nature of blockchain allows multiple parties to access and share data without relying on a centralized organization or intermediary. This characteristic proves particularly valuable in sectors like finance, healthcare and supply chain management, where secure and transparent data sharing is crucial. For example, all participants in a supply chain management system, including suppliers, manufacturers, distributors and retailers can access and share data seamlessly on the blockchain network [22].

Table 1.4 Blockchain-based storage and traditional cloud storage [21]

Blockchain-based storage	Traditional cloud storage
Blockchain storage is cheaper.	Cloud storage is more expensive.
More secure storage.	Less secure storage.
Uses different methodologies to encrypt data.	Uses virtual space where data can be accessed remotely.
Decentralized system.	Centralized system.
Faster access to data.	Slower than blockchain.
Users have access to smaller data chunks.	Users have access to entire files.
Data is immutable.	Data is mutable.
Blockchain can track data.	Cloud cannot track data.

Impact of blockchain on data sharing

Blockchain's decentralized approach is poised to revolutionize the way organizations share data in the near future, particularly in the context of Web 3.0. The key advantages that distinguish Web 3.0 from its predecessor (Web 2.0) include the absence of a central owner or regulators, heightened security and privacy measures, enhanced reliability, load balancing capabilities and reduced costs (Table 1.5).

Data removal challenges in blockchain

The immutability and tamper-proof nature of blockchain technology pose challenges to the process of data removal. Once data is added to a blockchain, it becomes practically impossible to alter or delete without obtaining consensus from the network's users.

Despite this inherent difficulty, a few methods exist for deleting data from a blockchain. One notable approach is through *Hard Forks*. A hard fork entails a permanent divergence in the rules that nodes in the network must follow. In the case of removing specific data, nodes can either continue using the original blockchain or opt for the new blockchain that excludes the targeted data. This results in the creation of two separate blockchains, with transactions from each blockchain no longer associated with the other.

Hard forks can be intentional or accidental, the latter often occurring due to bugs or errors in the blockchain. They are significant events in any blockchain community, typically well-discussed and planned if intentional. An example of a planned hard fork occurred on the Bitcoin platform, leading to the creation of a new blockchain known as *Bitcoin Cash Node* (BCHN), later considered the official Bitcoin Cash due to its substantial mining power [23–25].

Table 1.5 Data sharing using web 2.0 and web 3.0 [22]

Parameter	Web 2.0	Blockchain/Web 3.0
Ownership	Restricted	Distributed
Anonymity	Can be compromised	Protected
Transparency	Minimal	Higher
Security	Intermediate	Enhanced

Pruning in blockchain

Pruning in blockchain refers to the selective removal of unnecessary data branches, akin to removing unnecessary branches from a tree or plant. In blockchain technology, each transaction creates a new state based on the previous state, resulting in the accumulation of a large and unwieldy state over time. This poses challenges for storing and syncing full nodes, impacting scalability and resource requirements. state pruning is a technique employed to reduce the complexity and size of the blockchain's state, enhancing efficiency and performance.

Three types of pruning approaches are commonly used:

Partial State Pruning retains the most relevant and frequently accessed data.

Recursive State Pruning divides data into smaller chunks and performs pruning recursively, often starting with the oldest or least important data.

Sparse State Pruning involves creating a sparse state containing the minimum data necessary to recreate the current state of the blockchain [26]. As such, state pruning offers significant benefits, particularly for blockchains with a large number of users and transactions. It reduces hardware and energy costs by requiring fewer resources for full nodes to store and process the entire blockchain state.

Off-chain storage in blockchain

Off-chain storage involves keeping data off the blockchain while maintaining reference to it. This approach helps in managing blockchain data efficiently while allowing access when needed. Off-chain storage does not necessarily imply data exclusion from the blockchain; rather, it means that the data is not publicly accessible. Off-chain transactions are conducted using values outside the blockchain offering speed and cost advantages compared to on-chain transactions. Organizations often manage blockchain-related data through a combination of on-chain and off-chain storage methods.

Unstructured data, such as files, contracts, photos and papers, constitutes a significant portion of blockchain implementations. Off-chain blockchain storage becomes useful for handling this unstructured data efficiently. Storing heavy datasets off the main blockchain nodes in an off-chain storage database reduces the storage needed by each node and eases blockchain traffic.

It's essential to approach data removal from a blockchain with caution as the technology's main strengths lie in its immutability and transparency. Careful consideration is necessary to avoid damaging users' trust in the system [27, 28].

Data authentication in blockchain

Data authentication, the process of confirming the authenticity and integrity of data stored on a blockchain, ensures that the data remains unaltered. Various techniques, including digital signatures, cryptographic hashing and consensus algorithms, are employed to authenticate data in a blockchain. This crucial process guarantees the accuracy and immutability of each block's data.

In a blockchain network, every transaction undergoes validation by a network of computers, known as nodes. The transaction is added to the blockchain only after achieving consensus, with each node verifying its accuracy. This robust procedure ensures the realness and unchangeability of data on the blockchain [29].

The use of public and private keys is integral to authenticating data in the blockchain. Private keys sign and validate transactions, while public keys identify users on the network.

This dual-key system ensures that only authorized users can modify the data, preventing unauthorized access and guaranteeing data integrity. Once data is added to the network, its immutability is ensured by data authentication, providing transparency for anyone to verify the information on the blockchain.

One widely used method for data authentication in blockchain is *hashing*. A cryptographic hash function generates a unique hash value for each block, serving as a digital fingerprint of the block's information. Any alterations to the data will result in a new hash value, making it easy to detect any attempted manipulation.

Digital signatures are another commonly employed method in blockchain technology for data authentication. Each network user is assigned both a public key and a private key. When making a transaction, the sender's private key creates a digital signature verified using the sender's public key, ensuring the transaction's authenticity and integrity.

While blockchain technology provides a secure, transparent and efficient method for data authentication, it has drawbacks, such as potential slowness, especially for larger datasets. The verification process for every block added to the chain can take time, impacting transaction speed [30].

Data privacy: navigating challenges in blockchain

Data or information privacy revolves around handling private and sensitive data in a secure manner, encompassing the collection, storage, processing and handling of personal information. In the contemporary business landscape, where data holds immense significance, ensuring the protection and privacy of data becomes a paramount concern.

In a private blockchain, data privacy is relatively straightforward as participants are restricted and a regulatory protocol is enforced and maintained by a central authority. Contrastingly, in a public blockchain maintaining data privacy for participants becomes challenging due to the absence of a central authority overseeing privacy and regulations.

Recent trends in data privacy law

In recent years, the spotlight on data privacy has intensified globally, evolving into a substantial policy and legislative issue, particularly with the advancement of blockchain technology. Traditional data controllers, which determine the objectives and means of data processing, differ from the decentralized nature of blockchain technology. This shift challenges compliance with regulations like the GDPR (General Data Protection Regulation).

Key considerations and challenges include:

Blockchain Network Structure: The peer-to-peer network structure of blockchain contrasts with centralized businesses, impacting how data is controlled and interacts with service providers.

Compliance Obligations: Users of blockchain technology may find compliance with regulations, such as GDPR, challenging. Ensuring the legality of personal data processing, obtaining user consent and meeting legal criteria are complex tasks.

E-Privacy Regulation: While the GDPR regulates various aspects, the draft E-Privacy Regulation focuses on specific communications data. Similar to GDPR, it requires a legal basis for data processing.

Jurisdictional Differences: The United States and EU follow distinct approaches to data privacy, with laws like GDPR and California Consumer Privacy Act of 2018 (CCPA) relying on traditional methods. The clash between decentralized blockchain architecture and centralized privacy laws poses challenges.

The traditional centralized approach to data privacy, namely, controlling data collection and service provider relationships contradicts the decentralized peer-to-peer architecture. Navigating these challenges requires a careful balance between regulatory compliance and the innovative potential of blockchain technology [31].

Blockchain-based architecture for data privacy preservation

A blockchain is composed of a comprehensive list of transaction records, forming a chain of blocks akin to a traditional public ledger. Each block references its preceding block through a hash value known as the parent block. The genesis block, the initial block on a blockchain, lacks a parent block. A block consists of a block header and a block body, with the header encompassing elements like block version, parent block hash, Merkle tree root hash, timestamp, nBits and Nonce. This blockchain architecture is pivotal in the context of data privacy preservation.

Blockchain-based data security

The focal point of the blockchain-based architecture is data security, particularly in the realm of smart automobiles. Interconnected services in smart automobiles offer substantial advantages but also pose privacy and security risks, such as location tracking and remote vehicle hijacking. Emerging services, including dynamic automobile insurance rates, are wirelessly updated through remote software.

The authors assert the resilience of their proposed blockchain-based architecture against basic security attacks, leveraging the inherent properties of blockchain technology. However, to comprehend the privacy implications of blockchain's decentralized architecture, it is crucial to weigh the privacy gains against the costs. Assessing whether the benefits outweigh the privacy costs requires a nuanced understanding of the disclosure and identification of personal information. For an in-depth exploration of privacy considerations related to blockchain technology, refer to the dedicated privacy chapter "Chapter 4: Privacy Considerations in Blockchain Technology" [32].

Blockchain technology manages an extensive record of transactions enveloped within multiple layers of data security. This inherent security is attributed to the structure and rules governing blockchain systems.

Blockchain data is considered generally secure due to the multilayered security it employs. Any illicit attempt to alter the blockchain ledger necessitates breaching multiple security layers, traceable back to the potential hacker, resulting in a loss of access to the network.

The process of adding new transactions, known as minting, adheres to specific rules ensuring the security of the blockchain. Each block must have a unique address, creating a linked chain of data blocks with traceability to the first block. New transactions undergo validation, encryption and consensus by nodes before integration into the blockchain.

Common security-related issues

Blockchains face potential security risks, including DoS attacks, endpoint security vulnerabilities, code vulnerabilities and data protection concerns. Attacks like Border Gateway Protocol (BGP) hijacks, routing attacks, eclipse attacks and Erebus attacks pose threats to blockchain networks.

Endpoint threats such as malware and crypto jacking also present security challenges. Malware infections, often utilized for cryptocurrency mining, compromise millions of computers. Crypto jacking involves using a user's web browser to mine cryptocurrency without their knowledge.

Furthermore, criminals exploit blockchain vulnerabilities for ransomware attacks using Bitcoin for extortion. For example, the *WannaCry* ransomware attack in May 2017 targeted Windows systems, encrypting files and demanding Bitcoin as ransom.

The 51% vulnerability allows attackers to take control of the entire blockchain, manipulate information and perform a double-spending attack. Blockchain faces multiple security risks, including criminal activity, private key security, transaction privacy leakage and smart contract vulnerabilities.

Blockchain is susceptible to nine out of the ten risks from the Open Web Application Security Project (OWASP) top ten list encompassing injection, broken authentication, sensitive data exposure, broken access control, security misconfiguration, insecure deserialization and more [33–35].

Refer to the dedicated security chapter for an in-depth exploration of various blockchain attacks and mitigation techniques.

CONCLUSION AND RECOMMENDATIONS

To recap, currently the majority, if not all, transactions are facilitated through intermediaries like banks. Blockchain technology emerges as a transformative solution, eliminating the need for these third-party entities. Acting as a distributed and decentralized digital ledger, blockchain records transactions and stores data securely. It offers attributes such as security, persistence, decentralization, anonymity and data integrity in transactions.

Private blockchains operate within restricted networks, allowing only authorized groups to contribute data. On the other hand, public blockchains function openly in decentralized networks recording all transactions on a transparent ledger accessible to anyone. Permissionless blockchains, recognized as open blockchains, contrast with permissioned blockchains referred to as closed blockchains.

Experts identify five layers in blockchain technology: Infrastructure or Hardware layer, Data layer, Network layer, Consensus layer, and Application and Presentation layer. These layers are categorized as Layers 0 to 3, providing a comprehensive framework for understanding blockchain technology.

Research is ongoing to develop data types with minimal memory requirements, advancing beyond traditional standards. Two models, the Transaction Data Model and Account-based Model, are utilized in cryptocurrencies like Ethereum. Bitcoin employs the Merkle tree structure with hashing operations on nodes generating the Merkle root. Ethereum calculates the Merkle root using the Merkle Patricia tree, ensuring the verification of untampered transaction data and block headers.

Blockchain policies are shaped by consensus mechanisms and encoded in smart contracts. Principles are established to support and regulate blockchain applications, emphasizing the technology's ability to create an accurate, enduring and cost-effective database for recording Internet-based data exchanges.

Bitcoin introduced the concept of data ownership. Individuals now have greater control over their personal data, deciding what to share, and with whom and establishing permissions and access controls.

Cryptocurrencies using proof-of-work mining, while resource-intensive, remain an attractive prospect due to occasional profitability. Data archiving involves moving inactive data to ensure regulatory compliance, while data retention in blockchain specifies how long a block can store information.

Blockchain data organized into blocks connected by cryptography forms an immutable chain. Data removal from blockchain is challenging due to its tamper-proof nature, but methods like Hard Forks, Pruning and Off-chain storage offer potential solutions.

Data authentication ensures the accuracy of each block's data by validating transactions through a network of computers called nodes. Recent years have seen increased global attention to data privacy, growing alongside blockchain technology development. Despite its security benefits, blockchain faces potential risks such as denial-of-service (DoS) attacks, endpoint security issues, code vulnerabilities and data protection concerns.

In conclusion, blockchain technology, despite being relatively new, demonstrates its potential in diverse fields such as humanitarian aid, supply chain management and disaster relief. As technology advances and research expands, blockchain has the potential to become a fast, efficient, reliable and secure solution for various transactions and complex data storage in a transparent manner.

Data management best practices with blockchain technology

Utilize Blockchain Technology: Leverage blockchain technology to remove intermediaries, enhancing security, persistency, decentralization, anonymity and data integrity in transactions.

Develop a Comprehensive Understanding of Blockchain: Gain insights into various aspects of blockchain technology, including distinctions between private and public blockchains, layers (infrastructure, data, network, consensus, and application and presentation layers) and diverse blockchain models used in different cryptocurrencies.

Implement Permission-Based Systems: Ensure data security by employing permission-based systems, allowing only authorized users to access data.

Establish Clear Policies and Regulations: Formulate robust policies and regulations to support and govern blockchain applications effectively.

Take Control of Personal Data: Empower individuals with control over their personal data. Decide what to share, and with whom, and establish permissions and access controls.

Understand the Data Lifecycle on Blockchains: Recognize the significance of data archiving, retention and deletion within the blockchain technology context.

Initiate Efficient Transaction Tracking: Leverage blockchain's inherent capability to easily track and trace the history of transactions.

Implement an Appropriate Consensus Mechanism: Choose and implement an appropriate consensus mechanism tailored to the specific needs of your blockchain application.

Empower Individuals with Data Control: Recognize that blockchain technology provides individuals with control over their personal data, fostering empowerment.

Fully Understand the immutability and tamper-proof nature of blockchain technology: Be aware of methods for deleting data, including hard forks, pruning and off-chain storage.

Validate Every Transaction: Ensure data accuracy and prevent tampering by validating every transaction within the blockchain network.

Mitigate Potential Risks: Be vigilant about potential risks such as denial-of-service (DoS) attacks, endpoint security issues, code vulnerabilities and data protection. Implement measures to mitigate these risks effectively.

REFERENCES

[1] J. Yli-Huumo, D. Ko, S. Choi, S. Park and K. Smolander, "Where Is Current Research on Blockchain Technology?—A Systematic Review," *Plos One*, 2016.

[2] Z. Zheng, S. Xie, H.-N. Dai, C. Xiangping and H. Wang, "An Overview of Blockchain Technology: Architecture, Consensus, and Future Trends," *6th IEEE International Congress on Big Data*, 2017.

[3] S. Ghoash, P. K. Aithal, P. S. Saavedra and S. Aithal, "Blockchain Technology and Its Types—A Short Review," *International Journal of Applied Science and Engineering (IJASE)*, vol. 9, no. 2, pp. 189–200, 2021.

[4] R. Marar and H. Marar, "Hybrid Blockchain," *Jordanian Journal of Computers and Information Technology*, vol. 6, no. 4, 2020.

[5] E. Eliacik, "4 Types of Blockchain Explained – Dataconomy," 27 April 2022. [Online]. Available: https://dataconomy.com/2022/04/4-types-of-blockchain-explained/

[6] T. K. Sharma, "Permissioned and Permissionless Blockchains: A Comprehensive Guide," 03 November 2022. [Online]. Available: https://www.blockchain-council.org/blockchain/permissioned-and-permissionless-blockchains-a-comprehensive-guide/

[7] S. Solat, F. Naït-Abdesselam and P. Calvez, "Permissioned vs. Permissionless Blockchain: How and Why There Is Only One Right Choice," *Journal of Software*, vol. 16, no. 3, pp. 95–106, 2020.

[8] A. Hayes, "How Does Bitcoin Mining Work?" 21 November 2019. [Online]. Available: https://www.investopedia.com/tech/how-does-bitcoin-mining-work/

[9] S. Verma, "A Beginner's Guide to Understanding the Layers of Blockchain Technology," 25 May 2022. [Online]. Available: https://www.blockchain-council.org/blockchain/layers-of-blockchain-technology/

[10] Cointelegraph, "A Beginner's Guide to Understanding the Layers of Blockchain Technology," 19 April 2022. [Online]. Available: https://cointelegraph.com/blockchain-for-beginners/a-beginners-guide-to-understanding-the-layers-of-blockchain-technology

[11] Techtarget, "What Is Data Type?" 30 November 2016. [Online]. Available: https://www.techtarget.com/searchapparchitecture/definition/data-type

[12] J. L. X. L. Xiaojing Yang, "Research and Analysis of Blockchain Data," *Journal of Physics Conference Series*, 2019.

[13] Hackernoon, "Bitcoin UTXO vs Ethereum's Account-Based Blockchain Transactions: Explained Simply," 22 August 2021. [Online]. Available: https://hackernoon.com/bitcoin-utxo-vs-ethereums-account-based-blockchain-transactions-explained-simply-164x37f5

[14] Safe, "EOAs vs. Smart Contract Accounts," 2023. [Online]. Available: https://docs.gnosis-safe.io/learn/gnosis-safe/eoas-vs.-contract-accounts

[15] Horizen, "The UTXO vs Account Model," Horizen Academy 21 February 2023. [Online]. Available: https://www.horizen.io/academy/utxo-vs-account-model/

[16] D. C. Alan McQuinn, "A Policymaker's Guide to Blockchain," *Information Technology & Innovation Foundation*, 2019.

[17] Y. Kishimoto, K. Masuda, C. Dai and M. Yano, "Creation of Blockchain and a New Ecosystem," *Blockchain and Crypt Currency*, 2020.

[18] Y. Liu, Y. Zhang, Y. Yang and Y. Ma, "DOCS: A Data Ownership Confirmation Scheme for Distributed Data Trading," *Systems*, vol. 10, no. 6, p. 226, 2022.

[19] A. Galiev, N. Prokopyev, S. Ishmukhametov, E. Stolov, R. Latypov and I. Vlasov, "ARCHAIN: A Novel Blockchain Based Archival System," *IEEE*, pp. 84–89, 2018.

[20] E. Konacakli and E. Karaarslan, "Data Storage in the Decentralized World: Blockchain and Derivatives," arXiv preprint arXiv:2012.10253, 2020.

[21] P. K. Kollu, M. Saxena, K. Phasinam, T. Kassanuk and M. Jawarneh, "Blockchain Techniques for Secure Storage of Data in Cloud Environment," *Turkish Journal of Computer and Mathematics Education (TURCOMAT)*, vol. 12, no. 11, pp. 1515–1522, 2021.

[22] R. Khalid, M. Naz, N. Javaid, A. M. Qamar, F. A. Al-Zahrani, M. K. Afzal and M. Shafiq, "A Secure Data Sharing Platform Using Blockchain and Interplanetary File System," *Sustainability*, vol. 11, no. 24, p. 7054, 2019.

[23] N. C. Yiu, "An Overview of Forks and Coordination in Blockchain Development," *arXiv preprint arXiv:2102.10006*, 2021.

[24] Coinmarketcap, "Hard Fork (Blockchain)," 21 August 2021. [Online]. Available: https://coinmarketcap.com/alexandria/glossary/hard-fork-blockchain

[25] W. Vermaak, "What Is a Hard Fork?" 21 August 2021. [Online]. Available: https://coinmarketcap. com/alexandria/article/what-is-a-hard-fork

[26] C. Gardiner, "Exploring the Benefits and Challenges of State Pruning in Blockchain Systems," *Harmony*, 2022.

[27] R. Pinto, "On-Chain versus Off-Chain: The Perpetual Blockchain Governance Debate," 06 September 2019. [Online]. Available: https://www.forbes.com/sites/forbestechcouncil/2019/09/ 06/on-chain-versus-off-chain-the-perpetual-blockchain-governance-debate/?sh=76949ba91f5e

[28] Blockchainsentry, "What Is Off-Chain Storage in Blockchain," 14 September 2022. [Online]. Available: https://blockchainsentry.com/blog/what-is-off-chain-storage-in-blockchain/

[29] L. J. Lacombe, "10 Examples of Blockchain Authentication," 21 August 2021. [Online]. Available: https://www.bcdiploma.com/en/blog/10-exemples-usage-authentification-blockchain-2021-08-17

[30] U. Otuekong, R. Singh, S. Awan, Z. Pervez and K. Dahal, "Blockchain-Based Secure Authentication with Improved Performance for Fog Computing," *Sensors*, vol. 22, no. 22, p. 8969, 2022.

[31] P. Shah, D. Forester, M. Berberich, C. Raspe, and H. Mueller, "Blockchain Technology: Data Privacy Issues and Potential Mitigation Strategies," *Practical Law – Thomson Reuters*, 2019.

[32] F. Daneshgar, O. Ameri Sianaki and P. Guruwacharya, "Blockchain: A Research Framework for Data Security and Privacy," *Workshops of the International Conference on Advanced Information Networking and Applications*, pp. 966–974, 2019.

[33] A. Bylund, "Blockchain Security Defined," 09 June 2022. [Online]. Available: https://www.fool. com/investing/stock-market/market-sectors/financials/blockchain-stocks/blockchain-security/.

[34] H. Guo and X. Yu, "A Survey on Blockchain Technology and Its Security," *Blockchain: Research and Applications*, vol. 3, no. 2, p. 100067, 2022.

[35] X. Li, P. Jiang, T. Chen, X. Luo and Q. Wen, "A Survey on the Security of Blockchain Systems," *Future Generation Computer Systems*, vol. 107, pp. 841–853, 2020.

Chapter 2

Blockchain governance considerations

Alamgir Faruque and Tahziba Tabassum

BACKGROUND

Since the 1980s, the term "governance" has become prevalent across various contexts, reflecting its increased use. Governance, in a broad sense, pertains to decision-making and the exercise of authority with a focus on the individuals possessing these capabilities. It encompasses all forms of governing, be it by a government, market, network, family, tribe, formal or informal organization, or territory and can manifest through rules, standards, authority or language [1]. Participants or users engage with a system through its governance structure, a ubiquitous feature in nearly every societal framework often found in unexpected places. Effective governance fosters compliance with regulations and encourages efforts to enhance conditions for the collective benefit [2].

In the realm of Blockchain systems, governance plays a pivotal role in ensuring the right incentives, growth, success and the establishment of an ecosystem with protocols and guidelines. Blockchain governance, within this context, encompasses the procedures, regulations and processes upholding the protocol. Maintaining decentralization is imperative for the success of most blockchain projects making governance challenging [3]. Decentralized networks and platforms rely on dynamic governance systems to sustain their initiatives' real-world use without a centralized authority. Blockchain projects must implement governance capable of handling decision-making complexity to mitigate uncertainty, delays and costs. Regulatory, compliance and privacy requirements significantly influence governance decisions [4].

Many corporate challenges demand a high level of privacy and compliance due to stringent legal requirements. Consequently, the development of diverse blockchain network architectures becomes essential. Before embarking on a project, a robust legal framework should safeguard specific legal assets necessitating a thoughtful choice of blockchain network type. The legal system has become a shaping force in the blockchain economy. Two primary types of blockchains exist: public and private. A public blockchain is preferable when decision control is not a priority, as it is less susceptible to major governance changes. Conversely, a private network becomes the optimal choice when a company requires enhanced control over network governance and business operations and transactions. Allocating participants roles, responsibilities, decision rights, incentives, compliance requirements and making technical design and architecture decisions are integral aspects of blockchain governance [3].

DISCUSSION

Blockchain technology not only enhances business efficiency but also fosters a culture of trust within corporate governance [5]. Traditionally, the decision-making power in corporations has been concentrated in a board of directors. However, this setup often leaves shareholders

DOI: 10.1201/9781003462033-3

with limited access to corporate documents and forces them to trust the directors for accurate information. Corporate rules, such as those governing fiduciary obligations, aim to safeguard shareholders and encourage directors to act in the best interests of stakeholders. Yet, monitoring corporate boards is expensive, and mandated disclosures may not fully address the inherent "agency costs" associated with delegated management.

By adopting blockchain technology, corporations can establish trust between directors and shareholders, leading to a significant reduction in monitoring costs. Enhanced, secure communication and providing shareholders with permissioned access to real-time records can obviate the need for burdensome disclosure rules. Collaboration between shareholders and directors becomes more effective when demands and concerns can be conveyed directly and frequently to the board and management [6].

The shareholder voting process, often convoluted and unclear in large corporations, can be streamlined and made more accessible through blockchain technology. This enables shareholders to participate in corporate voting and witness the process in a transparent manner. Blockchain also extends transparency to internal corporate matters, reducing the burden and expense of the diligence process in various transactions. As parties represent their corporate affairs to one another in every arrangement, blockchain lessens the associated costs while fostering greater confidence between transactional parties. The increased efficiency in the diligence process could potentially encourage more transactions, as they become more cost-effective, and the risks linked with significant deals are mitigated [5].

Navigating blockchain regulatory challenges and compliance risks

As mentioned, the phenomenon of blockchain has garnered interest from private individuals, businesses, financial institutions and governments, concurrently attracting attention from criminal elements. While blockchain technology has the potential to reshape markets, it introduces a myriad of compliance questions and concerns. Currently, legal systems struggle to adequately regulate blockchain technology and its diverse applications, resulting in considerable uncertainty in the contemporary economic landscape.

In blockchain transactions, pseudonymity is a key feature, making them not entirely anonymous but lacking complete transparency. For instance, in Bitcoin, users receive a public address for transactions, visible to all due to obligatory documentation. However, these addresses are often challenging to link to their actual owners. A significant compliance issue arises from the absence of robust *Know Your Customer (KYC)* and anti-money laundering (AML) regulations. While Bitcoin's proof-of-work system addresses user security to some extent, the inherent lack of transparency in blockchain transactions creates opportunities for criminal activities, including money laundering, terrorism financing and corruption [7].

As with any transformative innovation, consensus on the regulation and architecture of blockchain-based systems is essential for widespread adoption. The cost of technology implementation can pose challenges for smaller companies, and technical barriers hinder rapid scaling of blockchain to handle high transaction volumes. As such, the implications of blockchain technology adoption must be considered for existing intermediaries and their stakeholders [5].

Regulatory responses vary globally. In the European Union (EU), the inclusion of Bitcoin trading platforms in the scope of the *Fifth Anti-Money Laundering/Counter-Terrorism Financing (AML/CTF) Directive* followed significant fraudulent activities in initial coin offerings (ICOs) [8]. *The General Data Protection Regulation* (GDPR) in the EU presents substantial challenges for businesses from technical, administrative and legal perspectives with potential fines reaching EUR 20 million [9].

Furthermore, some nations, such as India, prohibit businesses and financial institutions from using cryptocurrencies as an asset class; while in Canada, digital currencies are not legal tender but can be used for payments. The United Kingdom Financial Conduct Authority collaborates with the Treasury Committee to assess the risks and benefits of cryptocurrencies and blockchain technologies [10]. These diverse regulatory landscapes underscore the need for a comprehensive approach to address the evolving challenges associated with blockchain technology.

Use of blockchain in enterprise data governance

At the data level, data governance involves procedures ensuring formal management of data assets across a company. A data governance model sets guidelines for authority, administration and decision-making concerning company-generated or managed data [4]. Blockchain technology plays a pivotal role in data governance, often complementing and facilitating data governance regulations. Key aspects of using blockchain for data governance include:

Data Storage: Acting as ledgers, blockchains allow the preservation of all data in its current state.

Decentralized Data: Unlike centralized data storage models, blockchains distribute data across ledgers, eliminating the need for a centralized authority. This decentralized approach enhances security [11].

GDPR Compliance: Blockchain aligns with regulations such as the *General Data Protection Regulation* (GDPR), eliminating the need for a single data management authority. Its security features, including transparency and consistency, make it a preferred choice for data governance. It expedites data control in various scenarios [8, 9].

Permanent Record: Blockchain's unchangeable nature ensures the permanence of recorded information.

Democratic Setup: Through a democratic setup using the "proof-of-stake" (PoS) protocol, everyone with equal rights to the data enhances transparency.

Compliance: Blockchain's emphasis on transparency and privacy often aligns with international laws, alleviating concerns about legal violations.

Traceability: The ledger's comprehensive recording facilitates easy tracking, reducing the risk of misuse.

Consistency: Computer algorithms and operations ensure the consistency of stored data.

Smart Contracts: These predefined statements encoded in the blockchain network have gained significant attention. These statements, such as "If/What If," are unanimously agreed upon by all users enhancing efficiency. Increasing smart contract encryption complexity further fortifies the robust blockchain network and safeguards sensitive data within it [12].

In conclusion, the integration of blockchain into enterprise data governance offers solutions to traditional challenges, providing a decentralized, secure and transparent framework for managing and safeguarding data assets.

Blockchain-based governance principles

Influenced by the blockchain stack and its components, the administration of blockchain is intricately connected to the ecosystem of protocols and guidelines within blockchain networks. Users navigate this evolving ecosystem that adapts rapidly to meet their needs, emphasizing the user's benefit as the ultimate goal [2].

Some key concepts of blockchain-based governance are as follows:

Scalability: Traditional centralized political institutions, such as the state, bureaucracy and representative democracies, historically addressed scaling issues by facilitating interactions among diverse or dispersed groups. Blockchain governance aims to scale effectively by accommodating the continuously shifting requirements and demands of its evolving ecosystem [13].

Elimination of Single Point of Failure (SPOF): Hierarchical structures and top-down centralized coordination in organizations can be inefficient, lacking flexibility and responsiveness. Such systems, designed to address historical necessities, are prone to risks like lack of openness, corruption, regulatory capture, abuse of power and even a return to authoritarianism. Blockchain governance seeks to eliminate SPOF, concentrating power by promoting decentralization. This decentralization is fundamental for achieving political effectiveness, equality, transparency and freedom [13].

Distributed Architecture and Trust-by-Computation: The principle of "Code is law" reflects the shift from traditional centralized vertical authority to a decentralized paradigm facilitated by blockchain technology. Peer-to-peer protocols, cryptographically verified, enable global coordination without intermediaries. Blockchain introduces large-scale decentralization, minimizing the human element and transferring trust from human employees to open-source code. In this distributed architecture, where "code is law," no single entity controls the network, ensuring redundancy through concurrent data storage. This neutrality of code, distributed consensus and transaction auditability reduce or eliminate frictions and failures inherent in centralized decision-making processes. Blockchain enables the implementation of various decentralized governance models and services without governmental supervision [13].

As such, blockchain-based governance principles prioritize scalability, the elimination of Single Points of Failure and the adoption of distributed architecture with trust-by-computation. These principles aim to enhance effectiveness, transparency and freedom while minimizing risks associated with centralized structures [13, 14].

Roles and responsibilities in blockchain governance

In contrast to traditional organizational structures where governance is typically centralized, blockchain being a decentralized technology lacks a singular authority. Instead, blockchain governance relies on four core communities to collectively address various issues with the number of these communities varying based on the specific blockchain [15]. The four central communities are as follows:

Core Developers: Responsible for creating, managing and maintaining the fundamental code of the blockchain, core developers have the authority to add, modify or delete code, directly influencing the blockchain's functionality, impacting all users [2].

Node Operators: With the increasing user base, demands arise for new features in the blockchain platform. In response to such requests, network users collectively decide whether to incorporate suggested features. This decentralized decision-making process ensures the integration of only essential features without overwhelming the application, a crucial aspect of blockchain governance [16].

Token Owners: Users holding blockchain tokens actively participate in the blockchain's ecosystem. They engage in governance through voting rights when changes, including feature modifications and established prices, are proposed. Token holders, seen as investors, express their views by holding a substantial portion of the total token supply [2].

Blockchain Team: The term refers to a business or non-profit entity undertaking various responsibilities to administer the blockchain. While blockchain platforms are often developed, owned and managed by the same individuals, there are instances where a specific Blockchain organization oversees a particular initiative managing blockchain governance. Key responsibilities include securing funding for project growth, engaging in negotiations with users and developers and marketing and promoting the project. In many cases, obtaining financing for the project takes precedence over directly influencing the blockchain's features. The blockchain team plays a role similar to a marketing team conveying investor needs to more critical communities, such as developers and node operators [2].

In essence, blockchain governance relies on the collaborative efforts of these four communities, each contributing to the effective functioning and evolution of the blockchain ecosystem. The decentralized nature of decision-making ensures a balanced and inclusive approach to managing and enhancing blockchain technologies.

Governance considerations in blockchain technology

Blockchain governance is comprised of four key elements, which serve as fundamental components. Identifying and understanding these features simplifies the overall governance structure of blockchain [2].

Consensus Algorithm: A pivotal aspect of blockchain governance, the consensus algorithm serves as the governing protocol. Responsible for maintaining the security and integrity of the distributed system, it defines the collaborative process through which consensus is achieved among nodes. The algorithm also outlines the validation process for records within the blockchain. Various consensus methods are employed in blockchain technology.

Proof of Work (PoW): Utilized by Bitcoin, PoW aims to distribute accounting rights and rewards through a competition for hashing power among nodes. Nodes compete to solve a mathematical problem based on the data from the previous block. The first node to solve this challenge generates the next block and receives a Bitcoin reward. *Satoshi Nakamoto* introduced *HashCash*, a Bitcoin math puzzle where each block's hash is used to produce the hash of the subsequent block, creating an immutable chained record of blocks. PoW involves terms and calculations integral to its functioning.

Hashes: Refers to 64-digit hexadecimal encrypted integers that must be validated before a new block is introduced. With technological advancements, generating a hash for a large quantity of data now takes milliseconds. However, miners engage in the time-consuming task of estimating this hash through computational efforts.

Nonce: Representing "number used once," the nonce is a set of digits incorporated into the hash. Mining begins with a nonce set to zero; and, the mining program on a node attempts to solve the hash by generating a hash from openly accessible data.

Solving the hash

A hash is considered solved when it is less than the current network target, which is a hexadecimal result derived from a mathematical formula determining the mining challenge. The miner successfully cracking the hash is rewarded with the current amount on the Bitcoin network.

As a security measure, PoW ensures the linking of blocks in the blockchain, creating a chain inversely proportional to the computational burden. The longest chain is universally trusted by peers. To manipulate the blockchain, an entity must possess over 50% of the global hashing power, making interference economically unfeasible. This establishes PoW as an effective security mechanism for the blockchain [17].

Proof-of-stake in blockchain governance

While PoW is still a common blockchain consensus mechanism, PoS presents an alternative approach to address resource waste and enhance security [18].

Coin age, a concept in PoS, is calculated by multiplying a coin's value by the time since its creation. Nodes gain increased network rights based on the duration of coin retention, receiving specific rewards according to coin age. In the PPCoin system, mining is still essential for obtaining accounting rights.

PoS introduces a formula where *Proofhash*, a hash value incorporating the weight factor, unspent output value and the ambiguous sum of the present time, is constrained by the coin's age (*Proofhash < Coin Age * Target Formula*). Mining complexity and coin age share an inverse relationship in PoS, promoting longer coin-holding periods. As such, PoS addresses resource waste associated with PoW and enhances blockchain security. To attack the blockchain, potential attackers must accumulate a significant amount of coins and hold onto them for an extended period, significantly raising the difficulty of an attack. *Ethereum*, *PPCoin*, *Nxt*, and *BlackCoin* are some examples of currencies utilizing PoS, each employing a random algorithm when allocating accounting rights, considering node rights [17].

Other consensus algorithms

Delegated Proof of Stake (DPoS): In the early stages of Bitcoin design, Satoshi Nakamoto envisioned a system where anyone could mine using a CPU, providing equal influence to each node in decision-making. However, technological advancements and the increasing value of Bitcoin led to the creation of mining-specific devices, concentrating hashing power in the hands of those with the most mining equipment. Ordinary miners faced diminishing opportunities to create blocks [17].

DPoS Concept: Builds on the original PoS consensus model, addressing speed and security concerns. Often referred to as "democracy" in the blockchain, DPoS involves currency holders voting to choose a delegation, known as the block producer. DPoS transactions are faster than PoW, but drawbacks include reduced decentralization and security challenges. Delegates, chosen to vote on behalf of others, may be removed if they fail to represent constituents adequately. However, this system's reliance on elected delegates makes it vulnerable to attacks and diminishes blockchain decentralization [18].

Proof of Authority (PoA): PoA is a consensus algorithm used in blockchain, especially suitable for private networks. It speeds up transactions by using identification as a stake. Validators stake their reputations and identities instead of actual money, ensuring a good reputation through hard work and adherence to strict laws. A select number of nodes, chosen by the algorithm, validate transactions, create blocks and maintain the network. Validators use specialized software, reducing the need for constant monitoring [15].

Practical byzantine fault tolerance

Byzantine Fault Tolerance is an effective way to address transmission faults in distributed systems. Nevertheless, the early Byzantine system needs exponential operations. It wasn't until 1999 that the PBFT (Practical Byzantine Fault Tolerance) system was presented and the algorithm complexity was decreased to a polynomial level, considerably increasing efficiency. It consists of five states:

Request: A client transmits a request to the master server node, which assigns a timestamp to the request.

Pre-Prepare: The request communication is recorded and assigned an order number by the master server node. The following server nodes receive a pre-prepare message transmitted by the master node. Initial decision-making about whether to approve the request is made by the other server nodes.

Prepare: If a server node decides to approve the request, it broadcasts a prepare message to every other server node and receives the prepare messages from the other nodes. If the request is accepted by the majority of nodes after messages have been gathered, the system will enter the commit state.

Commit: Every node in the server transmits a commit message to every other node when it is in the commit state. At the same time, a server node may assume that the request has been accepted by the majority of nodes if it gets the maximum number of commit messages. The node then follows the directions contained in the request message.

Reply: Client requests are answered by server nodes. The request is sent again to the server nodes if the network latency prevents the client from receiving a response. The server nodes only need to transmit the reply message repeatedly if the request was successful [17].

In short, the various consensus methods have an impact on how users participate in blockchain governance as well as general public governance. For example, using PoW or PoS as a mechanism strengthens the position of users and miners and makes them crucial to the system's continuity, safety and relevance. The users are co-producers of blockchain governance in both of these systems. In the PoA, users are the system's primary beneficiaries, whereas some groups perform the functions of a public agency to produce public services. This difference is related to the beneficiary or client role of citizens in traditional public administration and new public management regimes, as well as the co-producer role of citizens in new public governance systems in the literature on public governance [15].

Risks and considerations in consensus protocols

The selection of a consensus protocol for a blockchain system comes with potential risks and considerations, impacting performance, system stability, incentives, data management and the governing structure.

Performance Risks: Inappropriately choosing a consensus mechanism can adversely affect the performance of nodes, the overall network speed and transaction processing. Optimal performance relies on selecting a consensus protocol aligned with the specific requirements and characteristics of the blockchain system.

System Failure: An unsuitable consensus method, especially for public blockchains, can lead to system disruptions. Nodes may cease functioning or encounter critical issues disrupting the entire network. The choice of consensus protocol should align with the intended use and scalability of the blockchain system to prevent system failures [19].

Incentives: Incentives play a crucial role in the operation of blockchain networks. Miners and other contributing entities expect rewards for their efforts, ensuring the health and efficiency of the network. An effective incentive system is essential for the sustained operation and growth of the blockchain ecosystem [2].

Information Management: Whether public or private, data is a fundamental component of blockchain networks. The decentralized nature of blockchain demands substantial data significantly impacting both on-chain and off-chain operations. This differs fundamentally from traditional governance approaches, often resulting in superior results but necessitating careful consideration of data management strategies.

Governing Structure: The governing structure of blockchain is more adaptable compared to traditional organization, often linked to the chosen consensus mechanism. Traditional governmental institutions and corporate identities tend to have well-defined and inflexible governing structures. In contrast, blockchain's governing framework needs to be fluid to accommodate the dynamic nature of the evolving network [2].

Blockchain governance strategies

In a decentralized blockchain network, achieving consensus among nodes is essential for the validation and security of data. Similarly, altering the regulations and processes within a network of stakeholder organizations requires consensus among these stakeholders. Governance mechanisms within decentralized networks and platforms continually evolve to facilitate decision-making on updates, roadmaps and dispute resolution in a fair and inclusive manner, as there is no central authority overseeing these processes. The blockchain governance model plays a pivotal role in defining the project's level of decentralization, accessibility, equity and its ability to balance the interests of diverse stakeholders.

Understanding how blockchain governance operates involves exploring various governance models. These models shed light on the intricate dynamics of decision-making within decentralized systems. Some of the most popular governance models include [1]:

Consensus among Stakeholders: One prevalent approach involves achieving consensus among stakeholder networks. This model emphasizes the importance of collective agreement among the involved parties to implement changes or resolve disputes within the blockchain network.

Decentralized Decision-Making: A crucial aspect of blockchain governance is the absence of a central authority. Decisions are made collectively through a decentralized process, ensuring a distributed and inclusive approach to managing the network.

Balancing Stakeholder Interests: Successful blockchain governance models strike a delicate balance between the interests of different stakeholders. This involves considering the diverse needs and preferences of participants to ensure an equitable and sustainable system.

Adaptive Mechanisms for Updates: The ever-evolving nature of blockchain technology necessitates adaptive governance mechanisms. These mechanisms enable the network to effectively incorporate updates and changes, ensuring its resilience and relevance over time.

As discussed, the governance strategies employed in blockchain networks are critical for their functionality, security and sustainability. By comprehending the diverse models shaping decision-making processes stakeholders can actively contribute to the evolution of decentralized systems.

Off-chain governance mechanisms

Off-chain governance mechanisms are often considered informal and discreet maintaining a balance among various end users, including regular users, miners, commercial entities and other community stakeholders. This governance approach relies on the collective acknowledgment and simultaneous execution of necessary updates and implementations by all involved parties within a blockchain ecosystem [20]. It enables external entities, not authorized by the protocol itself, to intervene in the blockchain protocol. In instances where consensus cannot be reached, the network may undergo a fork, resulting in two chains using different software versions, with the chain possessing the highest transactional hashing power considered the successor of the original chain [2].

In the off-chain model, protocol development takes place in diverse forums such as conferences, internet forums and mailing lists. Examining the informal, off-chain processes that influence decision-making reveals the sometimes problematic nature of these procedures. A notable example is the block size debate within the Bitcoin community, which revolved around the question of whether to increase block sizes to enhance transaction processing capacity. This debate unfolded across various off-chain platforms and events involving agreements, disagreements, in-person discussions at conferences and the threat and execution of blockchain forks. An illustrative outcome of this debate is the emergence of *Bitcoin Cash* as a separate blockchain, with nearly identical code having lost consensus with the main Bitcoin blockchain [14].

Despite its merits, off-chain governance has two notable downsides. First, registering an individual user's opinion can be challenging; and, second, the implementation of changes can be slow to take effect [14]. These drawbacks highlight the need for careful consideration and refinement of off-chain governance mechanisms to address potential challenges and enhance overall effectiveness.

Participants in off-chain governance

Off-chain governance involves a series of steps that commence with core developers proposing modifications to the protocol through official channels, such as *Bitcoin Improvement Proposals (BIPs)* and *Ethereum Improvement Proposals (EIPs)*. These proposals are regularly documented in the project's official repository, typically hosted on platforms like *GitHub* or *Microsoft*. Stakeholders, including core developers, miners and other interested parties, engage in private and public discussions to express approval or criticism of the proposed improvements.

Subsequently, core developers assess the willingness of node owners and miners to adopt the proposed software updates. In an ideal scenario, unanimous agreement is reached and the implementation of code changes proceeds seamlessly. Transparent communication, achieved by announcing changes in advance, plays a crucial role in keeping stakeholders well-informed throughout this process.

When stakeholders cannot reach a consensus, they face two options. Initially, they may attempt to persuade other involved parties to support their viewpoint. If consensus remains elusive, an alternative option is to initiate a *hard fork* of the protocol. This allows stakeholders to maintain or alter features they consider essential, resulting in a divergence of the blockchain into two separate paths [21].

In summary, the key participants in off-chain governance include core developers, stakeholders expressing their opinions and node owners and miners who play a pivotal role in the implementation of proposed protocol changes. The success of off-chain governance relies on effective communication, consensus-building efforts; and, when necessary, the ability to navigate and manage protocol forks.

Types of off-chain governance

In the realm of PoW blockchain systems, both Bitcoin and Ethereum employ off-chain governance mechanisms. Notably, significant updates to the Bitcoin protocol undergo active online discussions among relevant parties. The Bitcoin core development team conducts regular online meetings – open to the public – where crucial matters are deliberated. While owning Bitcoin provides a voice in off-chain discussions, it does not confer voting rights as Bitcoin lacks a governance token. Consequently, all communication and voting processes in the Bitcoin network occur off-chain.

It is important to note that despite the perceived political influence associated with the hash power of large mining pools in Bitcoin, miners do not possess unique governance rights. Their role is limited to approving or disapproving transaction blocks, and the distribution of Bitcoin mining power does not equate to a distribution of governance powers.

Similarly, Ethereum's governance structure closely mirrors that of Bitcoin, emphasizing off-chain communication and decision-making [21]. Both blockchain systems rely on open discussions and the collective input of stakeholders, demonstrating a shared commitment to decentralized governance principles.

On-chain governance mechanisms

On-chain blockchain governance occurs directly within the blockchain protocol, requiring the integration of proposed modifications or decision-making processes into the network code. In this governance model project developers or node controllers hold the authority to drive organizational decisions and cast votes on new initiatives. Unlike traditional systems where each node has one vote, on-chain governance often assigns voting power based on the amount of native coins held by each node. This collective decision-making approach aligns with decentralized governance models, eliminating the need for a single central body in charge.

End users typically lack a direct voice in project management as developers and nodes are rewarded for participating in blockchain voting procedures. While specific eligibility criteria for voting may vary across projects, a code-based system involving pertinent stakeholders remains a common feature. On-chain governance offers the advantage of rapid decision-making as proposed changes are promptly imported into the project code; and developers are incentivized to vote, especially when supported by voting and implementation deadlines.

However, challenges exist, such as potential low voter turnout and undemocratic decision-making when certain nodes choose not to vote. In such instances, nodes with larger currency holdings wield disproportionate influence over the project's future trajectory [22]. Despite these challenges, on-chain governance expands the pool of contributors to blockchain governance processes, which reduces the risk of chain splits and forks, and allows users to submit their input.

While on-chain governance brings transparency and inclusivity, certain aspects of decision-making may still occur outside the blockchain. Proposals are often discussed in various forums, blog entries and social media platforms before being subjected to the formal voting process, even though the eventual decisions may be encoded in smart contracts. Although on-chain governance has the potential to rejuvenate the concept of system administration, the development of a governance model can be time-consuming, especially when introducing new tools [14].

Participants in on-chain governance

On-chain governance systems exclusively operate online, distinguishing them from informal governance systems that involve both offline collaboration and online code modifications for implementing changes. In on-chain governance, code updates serve as proposals for alterations to a blockchain requiring developers to submit improvement requests before changes can be applied. The coordination and consensus-building responsibilities rest with a core group, predominantly comprised of developers. The key participants in on-chain governance typically include:

Miners: Responsible for running nodes and validating transactions on the blockchain.
Developers: In charge of the foundational blockchain algorithms, often holding roles as users or investors in various cryptocurrencies.
Users or Participants: Individuals who invest in and utilize different cryptocurrencies.

Encouraging participation in the process involves providing economic rewards to participants. For instance, nodes may receive a share of overall transaction fees in exchange for their voting activities and developers may be compensated through additional funding sources. It is important to note that not every node carries equal influence; nodes with a larger holding of coins have more voting power than those with a smaller holding.

The suggested changes undergo a voting process, and participants or nodes decide whether to accept or reject the proposed modification. If approved, the change is integrated into the blockchain. Conversely, if the proposed change is unsuccessful, the updated code may be rolled back to a version preceding the baseline. This dynamic reflects the influence disparity among nodes in on-chain governance implementations. In short, the participants in on-chain governance play distinct roles in the decision-making process, and the distribution of influence is often tied to the amount of cryptocurrency held by each node [23].

Types of on-chain governance

Various blockchains implement on-chain administration in distinct ways, showcasing diverse models for decentralized decision-making. Two notable examples are Tezos and DFINITY:

Tezos: Employs a self-amending blockchain, wherein proposed modifications to the blockchain are first implemented in a test version. If the changes prove effective, they are integrated into the production version of the blockchain; otherwise, they are rolled back. Tezos also emphasizes the importance of determining the precise number of votes needed for a change to pass. Notably, the network allows voting on the voting process itself, offering a unique approach to address inefficiencies arising from dominant stakeholders and low voter participation [23] [22].

DFINITY utilizes an algorithmic governance approach through its *Blockchain Nervous System* (BNS). *Neurons* serve as the basic unit for the BNS acting as the network's voting units. Neurons must select a proposal category before submitting an amendment proposal, with categories based on economic, policy or code considerations. Each proposal submission incurs a fee, acting as a natural quality check against irrelevant or inconsequential submissions. However, this fee requirement raises concerns about potential pay-to-play issues, where stakeholders with more resources may disproportionately influence decision-making solely based on their wealth. Additionally, the DFINITY governance structure includes an automated voting mechanism, allowing Neurons to duplicate other people's votes [22].

These examples help highlight the diverse approaches to on-chain governance each with its unique features and considerations. As discussed, while Tezos focuses on self-amendment and iterative changes, DFINITY adopts an algorithmic model, with specific voting units proposing a distinctive way to categorize and process amendments.

Forks and blockchain governance changes

Many blockchains operate on open-source code, allowing accessibility for review by interested parties. In cases where major decisions divide the community or when someone has an innovative concept, individuals have the option to fork the project. *Forking* involves utilizing the code from the original project to create something new, enabling users to establish competition, reject previous projects or implement significant changes through this type of governance. Users actively participate in administration when consensus protocols validate transactions, allowing miners and stakeholders to influence the approval of transactions and protocol updates.

Forks, central to both off-chain and on-chain governance, come in two types:

Hard Fork: A fundamental revision to a blockchain network's protocol rendering previously valid blocks and transactions invalid, or vice versa. This significant protocol change results in the network splitting into two branches – one adhering to the old protocol and the other following the new one. Developers may implement hard forks for reasons such as addressing critical security flaws, introducing new functionality or rolling back transactions. Regardless of the cryptocurrency platform, hard forks occur on any network using cryptography. Miners, who establish the rules governing the network's memory, need to agree on the new rules and standards for a legitimate block prompting a "fork" to signify a modification or deviation from the protocol [12].

Soft Fork: In contrast, a soft fork is a modification to the software protocol where only the majority of miners need to upgrade to apply the new rules. Unlike hard forks that necessitate an upgrade by all nodes and agreement on the new version, soft forks result in a single chain. While soft forks are a crucial tool for making significant changes to blockchain software, they may not be as exciting for investors because the final result is always a single chain. Investors do not receive any new, distinct chain that would reward them with additional assets through a soft fork [12].

Forks play a pivotal role in the governance of blockchains, facilitating changes to protocols and offering a mechanism for addressing diverse challenges within decentralized systems.

The DAO case study

A decentralized autonomous organization (DAO) is designed to formalize decision-making processes within an organization, eliminating the need for traditional paperwork and personnel while establishing a decentralized management system. The functioning of a DAO involves the following key steps:

Smart Contract Development: A group of individuals writes smart contracts, which are programs that power the organization.

Crowdsale (ICO): During a crowdsale or ICO, individuals can contribute funds to the DAO by purchasing coins that represent ownership, providing the necessary resources for its operation.

Commencement of DAO Operations: The DAO becomes operational once the funding period concludes.

Proposal and Voting: Participants can propose spending suggestions for the funds, and members who have invested can vote on these proposals [24].

Launched on April 30, 2016, The DAO was among the pioneering DAOs, designed to function as an investor-directed venture capital firm. By May 15, 2016, it had raised over $100 million, eventually surpassing $150 million from over 11,000 members and marking it as the largest crowdfunding project in history at that time. Despite its success, concerns were raised about the code's vulnerability during the crowdsale, indicating that the marketing was more effective than the implementation.

Following the crowdsale, discussions emerged regarding the necessity of addressing vulnerabilities before financing proposals. On June 12, one of The DAO developers, *Stephan Tual*, disclosed a bug in the program. Although no funds were lost at that time, it led to more than 50 project proposals for DAO token holders to vote on. Notably, the Ethereum network, supporting up to $1 billion worth of *ether*, remained secure and functional. The incident underscored the susceptibility of smart contracts to attacks emphasizing the importance of thorough testing and prompt patching, as demonstrated by maker DAO swiftly addressing a vulnerability during testing [25].

Regrettably, an unknown attacker exploited vulnerabilities during the efforts of coders to address issues within the DAO system. The DAO, operating on the Ethereum platform and funded through token sales, experienced a significant breach. By June 18th, the attacker successfully drained over 3.6 million ether into a "child DAO" that mirrored The DAO's architecture. This incident caused a notable drop in the value of ether, declining from over $20 to just under $13.

Efforts to prevent further theft included attempts to split the DAO, but the necessary votes were not secured in time. Compounding the situation, all the ether funds were consolidated in a single location, an outcome unforeseen by the project designers [25].

As the Ethereum community grappled with the attack, the potential failure of The DAO posed disastrous consequences for both the evolving Ethereum network and financial losses for participants. At that time, The DAO held a substantial portion, approximately 14%, of the circulating ether (ETH), magnifying the impact of the security breach. The incident posed an existential threat to the promising Ethereum community and technology, just one year into its development [24].

Response to the DAO hack

Initially, *Vitalik Buterin*, Ethereum's creator, proposed a soft fork for the Ethereum network, introducing a code segment to blacklist the perpetrator and prevent the movement of stolen funds. In response, the alleged attacker, or someone claiming their identity, released an open letter asserting the legal acquisition of funds based on smart contract guidelines.

Emotions heightened when the same alleged attacker, or an impersonator, threatened to thwart any soft fork by bribing Ethereum miners to refuse participation. This move divided the Ethereum network and raised questions not only about technical challenges but also about the philosophical and moral foundations of the technology and the leadership of the Ethereum initiative [25].

Despite a bug in the soft fork's code, a hard fork was proposed and eventually implemented after extensive debate. The hard fork involved transferring The DAO's ether to a different smart contract, allowing investors to retrieve their funds and effectively rolling back the Ethereum network's history to pre-DAO attack status. Given the supposed permanence and censorship-resistant nature of blockchains, this decision sparked controversy [24].

The hard fork faced uncertainties initially, requiring consensus from miners, exchanges and network administrators. Ratified on July 20, 2016, at block 192,000, the Ethereum hard fork was implemented despite some dissent within the community [25].

While the majority accepted the change, a faction opposing the hard fork led to the creation of *Ethereum Classic* (ETC), maintaining the original pre-forked version of Ethereum. The attacker, although the funds taken from The DAO were returned to investors, retained stolen tokens on the Ethereum Classic chain valued at approximately $8.5 million. The DAO hack and subsequent Ethereum split underscored significant concerns in the community and highlighted the decisions that ensured Ethereum's survival in its early stages. In hindsight, these choices positioned Ethereum as a crucial element in the blockchain, cryptocurrency and autonomous finance landscape [25].

CONCLUSIONS AND RECOMMENDATIONS

In summary, the evolving landscape of blockchain governance necessitates a forward-looking approach to establish standards, foster trust and adapt regulatory frameworks to unlock the full potential of this transformative technology.

Blockchain governance encompasses the integration of standards, culture, laws, codes, individuals and organizations that collectively facilitate coordination and shape a specific entity. It encapsulates motivations, rules, and activities, contributing to the decision-making process. As blockchain technology gains increasing significance in business and private markets, establishing more suitable governance becomes imperative for seamless operations. An international standard should ideally be instituted, serving as a guiding framework across jurisdictions. Currently, many businesses face limitations in leveraging blockchain due to a lack of regulatory clarity [7].

Blockchain technology redefines the traditional notion of governance and challenges existing societal foundations. The resilience of centralized institutions will be tested in a future moving toward greater individual freedom and increased stakeholder involvement. Even entities accustomed to centralized governance models benefit from checks and balances, fostering transparency and accountability. Creative mixed governance structures offer solutions to address these challenges [22].

Several countries, including Australia, Estonia and the United States, have initiated blockchain governance applications in corporate finance. The US state of Delaware stands out for embracing the transformative potential of blockchain in reshaping business practices. As a central hub for corporate law, Delaware sets the precedent for other states looking for guidance in the absence of their own regulatory frameworks [15].

While blockchain technology holds incredible promise, fostering trust requires effective governance. Regulatory approaches must be flexible to enable blockchain's disruptive impact on financial markets and organizations. Notably, blockchain's achievement lies in generating trust without intermediaries, driven by the immutability of blockchain data. Corporate governance has an opportunity to innovate, redefine market value and improve internal processes by incorporating blockchain into innovation programs and making it a fundamental part of enterprise architecture governance [26].

Best practices in blockchain governance

Implementing a well-thought-out governance model is a critical step for leaders planning to integrate blockchain systems into their operations. This model should define rules for controlling the blockchain network and making necessary adjustments. To ensure optimal results, the following best practices for blockchain governance are suggested [27]:

New Partner or User Addition Process: Clearly describes the process for adding new partners or users to the blockchain network. A well-defined onboarding procedure contributes to smooth network integration.

Orderly Removal of Bad Players: Outlines a procedure for removing undesirable participants from the network in an organized manner. This ensures the maintenance of a healthy and secure blockchain environment.

Individual Policy Management: Enables every user and party in the network to manage their policies within the blockchain governance paradigm. This promotes autonomy and customization for stakeholders.

Adaptability to Operational Changes: The governance model should exhibit flexibility to adapt to changes in the operational processes of the blockchain. This agility is essential for keeping the system relevant and effective over time.

Balancing Stakeholder Interests: Maintains a balance between the interests of all significant stakeholders within the blockchain network. Various implementations of a blockchain governance model exist, with viability dependent on project-specific variables. Some models prioritize community involvement, emphasizing inclusivity, while others value well-informed, professional opinions [22].

Clearly Defined Internal Rules and Accountabilities: Clearly states internal rules and accountabilities aligned with the organization's goals. This clarity ensures that the governance structure is effective and aligned with the overarching objectives.

Transparency Among Stakeholders: Fosters transparency among stakeholders to govern the system more efficiently. Open communication and visibility into decision-making processes contribute to trust and collaboration.

Regular Review and Evaluation: Regularly reviews and evaluates all governance-related processes in advance. This proactive approach ensures continuous improvement and adaptation to the evolving needs of the blockchain network.

Adhering to these best practices establishes a robust foundation for effective blockchain governance, enabling organizations to navigate the complexities of decentralized systems with clarity, flexibility and transparency.

REFERENCES

[1] Pelt, R. van, Jansen, S., Baars, D., & Overbeek, S. (2020). Defining blockchain governance: A framework for analysis and comparison. *Information Systems Management*, 38(1), 21–41. https://doi.org/10.1080/10580530.2020.1720046

[2] 101 Blockchains. (2022, August 15). Blockchain governance principles: Everything you need to know. *101 Blockchains*. Retrieved November 24, 2022, from https://101blockchains.com/blockchain-governance/

[3] What is governance in Blockchain? *FutureLearn*. (n.d.). Retrieved November 23, 2022, from https://www.futurelearn.com/info/courses/introduction-to-blockchain-dlt/0/steps/250310

[4] Covarrubias, J. Z. L., & Covarrubias, I. N. L. (2021, January 22). Different types of government and governance in the Blockchain. *Virtus InterPress*. Retrieved November 24, 2022, from https://virtusinterpress.org/Different-types-of-government-and-governance-in-the-blockchain.html

[5] Validation request. (n.d.). Retrieved December 7, 2022, from https://www.conference-board.org/blog/environmental-social-governance/Blockchain-Organizational-Governance-Impact

[6] Pizztao, L. (2021, August 18). Blockchain as a tool for corporate governance. *Cygnetise*. Retrieved December 6, 2022, from https://www.cygnetise.com/blog/blockchain-as-a-tool-for-corporate-governance

[7] Teichmann, F., & Falker, M.-C. (2020, November 20). Compliance risks of blockchain technology, decentralized cryptocurrencies, and stablecoins. *zur Startseite*. Retrieved April 5, 2023, from https://ul.qucosa.de/landing-page/?tx_dlf%5Bid%5D= https%3A%2F%2Ful.qucosa.de%2Fapi%2Fqucosa%253A72845%2Fmets

[8] Gikay, A. A. (2019). European consumer law and Blockchain-based financial services: A functional approach against the rhetoric of regulatory uncertainty. *Tilburg Law Review*, 24(1), 27–48. Retrieved October 10, 2022, from https://doi.org/10.5334/tilr.135

[9] Piao, Y., Ye, K., & Cui, X. (2021). A data sharing scheme for GDPR-compliance based on consortium blockchain. *Future Internet*, 13(8), 217. Retrieved October 10, 2022, from https://doi.org/10.3390/fi13080217

[10] Cumming, D. J., Johan, S., & Pant, A. (2019). Regulation of the crypto-economy: Managing risks, challenges, and regulatory uncertainty. *Journal of Risk and Financial Management*, 12(3), 126. Retrieved October 10, 2022, from https://doi.org/10.3390/jrfm12030126

[11] AdminDataQg-2022. (2022, January 8). Blockchain: An accelerator for Data Governance. *DataQG*. Retrieved December 8, 2022, from https://dataqg.com/qgblogs/blockchain-an-accelerator-for-data-governance/

[12] Frankenfield, J. (2022, October 10). Hard fork: What it is in blockchain, how it works, why it happens. *Investopedia*. Retrieved November 24, 2022, from https://www.investopedia.com/terms/h/hard-fork.asp

[13] Atzori, M. (2016, January 2). Blockchain technology and decentralized governance: Is the state still necessary? *SSRN*. Retrieved October 20, 2022, from https://papers.ssrn.com/sol3/papers.cfm?abstract_id=2709713

[14] Blockchain governance on Decentralized Networks. *Gemini*. (n.d.). Retrieved November 23, 2022, from https://www.gemini.com/cryptopedia/blockchain-governance-mechanisms

[15] Tan, E., Mahula, S., & Crompvoets, J. (2022). Blockchain governance in the Public Sector: A Conceptual Framework for Public Management. *Government Information Quarterly*, 39(1), 101625. https://doi.org/10.1016/j.giq.2021.101625

[16] Blockchain governance, described!! *Blockchain Simplified*. (n.d.). Retrieved March 8, 2023, from https://blockchainsimplified.com/blog/blockchain-governance-described/

[17] Mingxiao, D., Xiaofeng, M., Zhe, Z., Xiangwei, W., & Qijun, C. (2017). A review on consensus algorithm of Blockchain. *2017 IEEE International Conference on Systems, Man, and Cybernetics (SMC)*. https://doi.org/10.1109/smc.2017.8123011

[18] Skh Saad, S. M., & Raja Mohd Radzi, R. Z. (2020). Comparative review of the blockchain consensus algorithm between proof of stake (POS) and delegated proof of stake (DPOS). *International Journal of Innovative Computing*, 10(2). https://doi.org/10.11113/ijic.v10n2.272

[19] Frankenfield, J. (2022, February 8). What is on-chain governance? *Investopedia*. Retrieved November 24, 2022, from https://www.investopedia.com/terms/o/onchain-governance.asp

[20] Reijers, W., Wuisman, I., Mannan, M., & De Filippi, P. (2018). Now the code runs itself: On-chain and off-chain governance of Blockchain Technologies. *SSRN Electronic Journal*. Retrieved October 10, 2022, from https://doi.org/10.2139/ssrn.3340056

[21] CoinMarketCap. (2021, November 18). Off-chain governance: CoinMarketCap. *CoinMarketCap Alexandria*. Retrieved December 6, 2022, from https://coinmarketcap.com/alexandria/glossary/off-chain-governance

[22] Smits, W.-J. (2019, March 14). Blockchain governance: What is it, what types are there and how does it work in practice? *Watsonlaw*. Retrieved April 4, 2023, from https://watsonlaw.nl/en/blockchain-governance-what-is-it-what-types-are-there-and-how-does-it-work-in-practice/

[23] Blockchain as a platform for Data Governance 2.0. *The Data School*. (n.d.). Retrieved October 20, 2022, from https://dataschool.com/data-conversations/blockchain-as-a-platform-for-data-governance-2-0/

[24] Siegel, D. (n.d.). Coindesk: Bitcoin, Ethereum, crypto news and Price Data. *CoinDesk Latest Headlines RSS*. Retrieved November 24, 2022, from https://www.coindesk.com/learn/2016/06/25/understanding-the-dao-attack/

[25] The dao: What was the dao hack? *Gemini*. (n.d.). Retrieved December 8, 2022, from https://www.gemini.com/cryptopedia/the-dao-hack-makerdao

[26] By: Michelle Bernier, By: Bernier, M., Related Content, By: & Scott Sumner, Sumner, S., By: & Pierre Lemieux, Lemieux, P., Corcoran, K., Rugy, V. de, & Donway, W. (2022, June 19). Shareholders' activism: The impact of Blockchain in corporate governance. *Econlib*. Retrieved December 6, 2022, from https://www.econlib.org/shareholders-activism-the-impact-of-blockchain-in-corporate-governance/

[27] Top 5 blockchain best practices to consider. *AIMultiple*. (n.d.). Retrieved April 5, 2023, from https://research.aimultiple.com/blockchain-best-practices/

Chapter 3

Operational risks in blockchain technology

Aditya Jayeshkumar Bhatt and Kesha Sisodia

BACKGROUND

Blockchain has expanded significantly beyond its cryptocurrency roots to infiltrate various businesses. Numerous sectors, including finance, healthcare, manufacturing and education, use blockchain applications to exploit this technology's unique characteristics. Data is stored in a distributed ledger using blockchain. Participants in the blockchain network can create, read and verify transactions stored in a distributed ledger, thanks to the integrity and availability provided by blockchain technology [1]. On the other hand, a blockchain forbids deleting and modifying any transactions or other data kept in its ledger. A digital signature, hash functions and various cryptographic mapping and protocols help to maintain and secure the blockchain system. Hash functions and digital signatures ensure that the transactions added to the ledger are non-repudiable, authentic and integrity-protected. In addition, as a distributed network, blockchain technology also requires a consensus mechanism which is essentially a set of guidelines to allow all members to agree on a single record [2]. Blockchain security merits further study because it is decentralized without involving a third party and requires building trust into a trustless architecture. Blockchain has specific operation challenges affecting organizational ability [3, 4].

DISCUSSION

Operational risk is the risk of suffering losses caused by people, insufficient or failing internal systems and processes or uncontrollable outside events. Every new technology has pros and cons that must be recognized and controlled. This is especially true when the technology represents a fundamental component of the organization's underlying IT infrastructure rather than just an overlaying application, as is frequently the case with blockchain [5]. Implementing blockchain-based applications requires handling several operational risks, including people–process–system risks.

People risk in blockchain technology

People risk refers to the possibility of momentary losses and poor social outcomes resulting from deficiencies in human capital and human resource management. Blockchain shared and distributed structure can benefit any business operations that demand a trusted environment to conduct secure and tamperproof transactions. However, companies can only enforce blockchain solutions if they have qualified developers with specific professional expertise to develop their network and applications [6]. People with highly technical occupations and creative minds behind design work are individuals that firms become unduly reliant on. If

DOI: 10.1201/9781003462033-4

institutions don't take the essential steps to be well-prepared, the businesses will suffer losses as soon as they leave. Blockchain developers are often considered as a *Talent Person Risk* for organizations. Blockchain developers frequently wear many hats depending on their enterprises' scale and specific situations. However, many blockchain developers fall under the categories of core blockchain and blockchain software developers. As such, people risk can create long-term risks for organizations [7]. According to a study conducted by *LinkedIn* and *OKX* utilizing data gathered from 180 nations throughout January 2019 and June 2022, there has been a mismatch between the number of qualified candidates available and the amount of talent needed in the global blockchain industry. According to the report, LinkedIn users in the blockchain industry increased by 76% from January to June. However, talent growth has been declining in the nations that create the majority of blockchain professionals [8]. *Melbin Thomas*, co-founder of *Sahicoin*, stated the existence of a "huge movement" of top talent from Silicon Valley's largest employers (Google, Facebook, Amazon) into the Web3 ecosystem because of the rich compensation packages and intriguing chances. According to industry experts, the sector faces a scarcity of knowledgeable people with the necessary skill sets and most of the newly hired talent learns on the job, which typically take about a quarter of their time [9].

Process risk in blockchain technology

Process risk refers to the possibility of monetary losses and poor social outcomes resulting from internal business processes that fail in various business aspects. While inefficient processes may be successful in achieving goals, they neglect to account for high expenses incurred, thereby impeding the achievement of the organization's goals. The critical scalability problem in the public blockchain is intensifying due to the widespread adoption of cryptocurrencies [10]. An anonymous user discovered an intelligent contract flaw in the library code used as a standard component of all multi-signature wallets *Parity Technology* offered in November 2017. The user assumed ownership of the library contracts. Next, the component was deleted, freezing the cash in 587 wallets, including more than $500,000 in Ether and other tokens. The Ethereum DEV team, Parity technologies and other community members designed and tested the first multi-signature wallet. The code underwent a thorough peer review and lacked known security flaws. Developers investigated the problem and found that double equals operators (= =) were being used instead of single equal (=). In contrast to the single equals operator, which sets the first parameter equal to the parameter or value following the equals, the double equals comparison operator returns a true or false value. This made it possible to defeat serial code validation. The attacker immediately sent the fabricated spent transactions to an Altcoin exchange address. It was very difficult to understand what was happening because more than 60 accounts were being used. This instance emphasizes the value of regular code reviews, internal and external audits. Automated checking techniques may also be deployed to find exploits [11].

As in the public blockchain, scalability is not a singular term. The effectiveness/performance of the consensus framework affects transaction throughput, latency and computational energy. The throughput can be directly increased by storing more transactions in a larger block, but the block propagation time will also increase. To demonstrate the restrictions on the number of transactions stored in each block, consider Bitcoin, where the block interval is approximately ten minutes, and the block size is about 1 MB [10]. In the context of blockchain, *Latency* is the amount of time required to perform a transaction, calculated from the time an input is received until the transaction is complete at the output [12].

To verify and process a transaction in a permissionless blockchain, thousands of nodes must reach a consensus. The fact that transactions are buffered while they wait to be confirmed logically results in higher latency. Every node in a public blockchain verifies and archives every transaction to maintain data security and integrity. *The number of nodes* means the number of entities connected to the blockchain network. As there are more nodes, there are more transactions, and more transactions are participating in the consensus process. The amount of computational energy also rises as the nodes multiply. Partial nodes (known as lightweight nodes) and full nodes are part of public blockchains. The full nodes are wholly necessary for the processing of partial nodes [12].

Table 3.1 illustrates information about three important parameters such as a Vulnerability ID, CVE ID and CVSS Severity. *Common Vulnerabilities and Exposure* (CVE) is a dictionary that categorizes vulnerabilities. The glossary assesses vulnerabilities, scores them using the Common Vulnerabilities Scoring System (CVSS) and then determines how dangerous they are. A CVE score is frequently used to rank the importance of vulnerabilities' security. *The Security Content Automation Protocol* is used to identify and classify vulnerabilities (SCAP). SCAP analyses vulnerability data and gives each vulnerability a special identification number. The CVSS, also referred to as the CVE score, is one of the various methods for evaluating the severity of vulnerabilities. The CVSS is a free set of guidelines that are used to evaluate vulnerabilities and rate their seriousness from 0 to 10. CVSS is currently at version 3.1 [13].

Computational Energy/Cost means the overall expense related to transaction verification. The mining process is connected to the consensus mechanism in the permissionless blockchain. Bitcoin employs the PoW consensus technique, necessitating specialized block-mining hardware. The high-end hardware's increased energy consumption may impact a public blockchain's implementation scale, resulting in higher computing costs [10]. *Block Size* in a blockchain such as Bitcoin is about 1 MB. The number of transactions that can be stored is thought to be minimal. Larger blocks can include more transactions, directly improving throughput and reducing latency. However, larger blocks would result in longer propagation times because heavier blocks take longer to transfer over a network. *The Transaction Cost* is pivotal in a public blockchain, such as Bitcoin. Based on the fees involved, miners frequently choose which transaction to verify. This directly influences transaction confirmation time affecting throughput and latency. The consensus process promises to compensate miners depending on the associated fees in a transaction with a small fee that may experience significant conformation delay [10]. Furthermore, processing more transactions in a block would require more CPU power [14].

System risk in blockchain technology

System risks refer to the risk of an entire system and its interoperability failing rather than just individual components failing. Integrating blockchain technology in various sectors has potential system challenges such as interoperability, security, standardization and storage requirements. For instance, interoperability, as in the healthcare sector, refers to data sharing across the entire blockchain network. Based on the vast array of providers and their extensive open presence, it is the leading cause for concern. Several actors could exist, including private doctors, physicians, insurance institutions and hospitals. Ensuring suitable interoperability across various institutions in healthcare can be time-consuming. Even if the use of blockchain technology is increasing, not all firms are adopting it because it doesn't improve their processes or because there aren't enough restrictions. But for governments, which must improve management and administrative activities, requirements like massive data storage, digitalization, efficiency, as well as security and privacy are crucial [2].

Since blockchain data is inherently decentralized, public–private data are often scattered throughout the symmetric public ledger, which could result in privacy leaks. Blockchain provides a setting where individuals can safely communicate information with others they

Table 3.1 CVSS coding flaws report

Vuln. ID	Summary	CVSS Severity
CVE-2018-11687	An integer overflow in the distributed BTR function of smart contract implementation for Bitcoin Red (BTCR), an Ethereum ERC20 token, allows the owner to perform an unauthorized increase of digital assets by providing a larger address [ ] array, which was exploited in the wild in May 2018, dubbed the "owneUnderflow" issue.	CVSS Version 3.x: 7.5 HIGH CVSS Version 2.0: 5.0 MEDIUM
CVE-2018-13485	The mintToken function of smart contract implantation for BitcoinAgileToken, an Ethereum token, has an integer overflow, allowing the contract's owner to set an arbitrary user's balance to any value.	CVSS Version 3.x: 7.5 HIGH CVSS Version 2.0: 5.0 MEDIUM
CVE-2016-10725	A non-final alert in Bitcoin Core prior to v0.13.0 can prevent the unique "final alarm" (intended to supersede all other alerts) because activities take place out of sequence. In the remote network alert system, the same behavior takes place (deprecated since Q1 2016). This has an impact on other uses of the codebase, including numerous altcoins and earlier versions of Bitcoin Knots (v0.13.0. knots20160814).	CVSS Version 3.x: 7.5 HIGH CVSS Version 2.0: 5.0 MEDIUM
CVE-2016-10724	Because of an infinitely sized map, Bitcoin Core prior to v0.13.0 allows denial of service (memory exhaustion) triggered by the remote network alert system (deprecated since Q1 2016) if an attacker can sign a message with a specific private key that was known by unintended actors. This has an impact on other uses of the codebase, including Bitcoin Knots prior to v0.13.0 knots20160814 and many altcoins.	CVSS Version 3.x: 7.5 HIGH CVSS Version 2.0: 7.8 MEDIUM
CVE-2018-10831	Prior to 2018-04-05, Z-NOMP had an incorrect Equihash solution verifier, allowing attackers to spoof mining share, as demonstrated by providing a solution with (x1 = 1, x2 = 1, x3 = 1...., x512 = 1) to bypass this verifier for any block header. This initially affected the Bitcoin Gold and Zcash cryptocurrencies and was still being used against smaller cryptocurrencies in May 2018.	CVSS Version 3.x: 7.5 HIGH CVSS Version 2.0: 5.0 MEDIUM
CVE-2018-6862	A profile field allows for Cross-Site Scripting (CSS) in PHP Scripts Mall Bitcoin MLM Software 1.0.2.	CVSS Version 3.x: 5.4 MEDIUM CVSS Version 2.0: 3.5 Low
CVE-2018-1000022	Electrum Technologies GmbH is a German company. Prior to version 3.0.5 of the Electrum Bitcoin Wallet, there was a Missing Authorization vulnerability in the JSONRPC interface that could result in Bitcoin theft if the user's wallet was not password protected. This appears to be an exploitable attack. The victim must go to a website that contains specially crafted JavaScript. This flaw appears to have been addressed in 3.0.5.	CVSS Version 3.x: 5.3 MEDIUM CVSS Version 2.0: 2.6 Low

Source: [13]

know and trust. However, in certain circumstances, such objectives could fall short particularly if those who access such data start acting maliciously. Due to security concerns, most blockchain-based organizations' clients or specific healthcare patients may experience extreme discomfort while disclosing or sharing their medical information. All data entered into the systems is permanently stored, which is one of the blockchain features. However, data protection regulations such as General Data Protection Regulation (GDPR) require that every user has both the right to be forgotten and the right to have their data erased. Hence, network storage data privacy requires specific consideration. While utilizing various data mining or big data analytics techniques, data privacy should be secured [2].

Blockchain technology transactional issues

According to recent research, 4.6 transactions are completed every second on Bitcoin. *Visa* typically completes 1,700 transactions every second. As such, Bitcoin transaction times are significantly slower in comparison. In 2018, a study by *Tata Communications* revealed that 44% of the firms surveyed were using blockchain, but it also hinted at the common issues of implementing the new technology. Scalability is an architectural issue developing as a barrier to blockchain uptake and useful applications. The block size is 1MB (1,048,576 bytes; however, through *SegWit*, that size can scale up to a theoretical 4MB). The number of transactions per block equals Block Size in Bytes divided by the Average Transaction Size in Bytes (1048576/380.04 = 2,759.12).

Transaction and *scalability* problems go hand in hand. One cannot simply boost scalability by altering blockchain constraints since the issue of scalability is packaged with blockchain value propositions. Here, the blockchain value proposition refers to two variables to increase Transaction Per Second (TPS). A variable is the *block size* (B), which is currently hardcoded at 1MB. Ideally, block size should be increased to increase TPS. The other variable is *block generation time* (TB), which is adjusted by changing the hashing puzzle complexity. The ideal solution would be to reduce TB in order to increase TPS [15].

Due to a design issue in Bitcoin, transactions can be changed after they are produced but before they are integrated into a block. This is known as *Transaction Malleability*. Changing the source and destination addresses or the transaction value is impossible. Still, changing other aspects of the transaction is possible, leading to a *transaction ID* (TXID) different from the original. *Mt. Gox* is one example of transaction malleability. The attacker could withdraw money fraudulently because Mt. Gox could not confirm the transaction's legitimacy because of the altered TXID [13].

System control risks

Hashes and public key encryption utilized for transaction signing are the two cryptographic primitives that make up the foundation of blockchain. The hash algorithm used by the majority of blockchains is *SHA-256*, which requires 2^{128} operation on a quantum computer to break using *Grover's approach*. While this renders SHA-256 robust to quantum attacks, the public key encryption algorithms may use do not have this property. When a sufficiently powerful quantum computer is created, the Elliptic Curve Digital Signature Algorithm (ECDSA) will be compromised, making practically all blockchains vulnerable. Although blockchain offers many advantages for managing and storing private data, it also has several drawbacks. Since information is maintained as a public ledger, blockchains often still have issues with privacy and secrecy. Various technologies based on anonymization or encryption can be used to safeguard the information's confidentiality. Transactional privacy is a well-known issue with blockchain data privacy. Many companies and people are worried about how transaction and SC operations, which spread over the network, might be traced. One of the most challenging difficulties with blockchain technology is transactional privacy. As a result, a number of techniques such as *mixing services* and *zero-knowledge proof* have been suggested to enhance blockchain anonymity. By moving money from the N input address to M output addresses while using mixing services, the goal is to provide transactional anonymity while preventing consumers from continually using the same address. Examples of such techniques in practice include *Mixcoin*, which is also capable of detecting dishonest transaction behaviors, and *Coinjoin* or *CoinShuffle* which shuffle output addresses via a third party [16].

Wallet risks

Cryptocurrencies don't exist in the same manner as traditional (fiat) currencies, like US dollars or UK pound sterling, as actual coins or paper currency are kept in pockets or purses. With cryptocurrencies a private or public key, or a combination of both, secure these assets. The private key is encoded with *Base58* to create a *Wallet Import Format* private key, which is a way to make a private key considerably easier to copy. The SHA-256 and RIPEMD-160 hashing algorithms are used to have the public key hashed. *RIPEMD* is used because it produces the shortest hashes, whose uniqueness is still sufficiently assured. This allows Bitcoin addresses to be shorter. Bitcoin's use of a hash of a public key might lead to unique characteristics due to the interaction between RIPEMD and ECDSA (the public key signature algorithm). However, numerous attack methods have the potential to make this crypto asset more vulnerable. A hardware wallet is a relatively new concept that strengthens the security of a user's Bitcoin holdings by storing a private key in *cold storage* offline. The hardware wallet does the transaction signing, and if there are no vulnerabilities, the transaction is kept secure in the user's wallet. An attack may isolate a specific cluster of nodes from a particular network. Connections between the target set of nodes and other nodes in the network can be rerouted and disabled by the attacker. Side-channel analysis and employing various testing frameworks can outline various issues ranging from insecure web interface, to lack of transport encryption, to lack of firmware updates or lack of data leakage testing [17].

Consensus risk

The consensus algorithm is the foundation of blockchain technology. Depending on the platform organization used to create it, multiple consensus methods for the validation of transactions have been employed in blockchain technology. These algorithms are guidelines to ensure the system operates correctly [18]. Design and modeling errors cause software failure, and such mistakes can lead to crashes and omissions. Omission failure is a model that defines a specific unit of time T, and all messages must be delivered at time T later. If it does not arrive in this time frame, it is assumed never arrived. It represents a possible problem that occurs within a network. However, Omission failures are caused by transmission problems like transmitter malfunctions, collision or buffer overflows.

Table 3.2 shows a comparison between various consensus algorithms and their properties. The three fundamental performance factors that must be considered for each consensus method are latency, throughput and bandwidth. The *Sybil attack*, in which a single attacker takes control of several network nodes by setting up various IP addresses, virtual machines, and user identities, is particularly dangerous for *Ripple*; a blockchain-based digital payment network and protocol that has its own cryptocurrency, *XRP*. Instead of mining blockchain, Ripple relies on a consensus mechanism via a group of bank-owned servers) [19, 20].

Table 3.2 Comparison of consensus algorithms

Consensus algorithm	Network	Cryptocurrency used	Throughput	Time required to validate the transaction	Pros	Cons
PoW	Permissionless	Bitcoin, Litecoin	Low	Large	Complex to validate but easy to verify. More secure	Waste of resources. Prone to 51% attack
PoS	Both	NXT, DASH, NEO	Low	Low	Threat of 51% attack is reduced. Greater Energy consumption, resources, and more effectiveness	Nodes with a large supply of collateral coins can take control of the network. Reducing the decentralization of the system
DPoS	Both	Cardano (ADA), EOS	Improved than PoS	Less than PoW	Fast, efficient, double spending attack risk get reduced	51% attack, Prone to DPoS
PBFT	Permissioned	Hyperledger, Chain	High	Less than PoW	Energy Efficient	Less Scalability
PoET	Permissioned	Hyperledger Solution	High	Less than PoW	Secure	Specialized hardware required
PoA	Permissionless	Decreed	Greater than PoW	Short	Increased speed of validation	Knowing the validator may cause interference
PoB	Permissionless	Slimcoin	Greater than PoW	Short	Secure distributed network	Wastage of resources

Source: [18]

Security failures

Blockchain technology has the power to change the way organizations conduct business; but just like any new technology, it also carries some hazards. The DAO (also discussed in Chapter 2) was an Ethereum blockchain-based platform for managing digital assets. On May 15, 2016, hackers extracted about $60 million worth of ETH in a single transaction [21].

Hacker Risks: The operational area of blockchain technology can be hacked in various ways, and each can have a tremendous effect on the growth of this technology. Software hacking means that the hacker may add fraudulent entries to the blockchain by manipulating the software that generates new blocks, defrauding other people of their just benefits. The network might suffer significant impacts, possibly even collapsing. Node Jacking means that hackers accessing many nodes could use them to alter network transactions or prevent others from connecting. A hacker may severely harm the overall security of the blockchain system. The exchanges that power Bitcoin are frequent targets for hackers, as with cryptocurrency exchanges. Customers unaware of the vulnerability in their accounts can have their cryptocurrency accounts stolen if a hacker has access to client data on an exchange. It can be kept in various wallets, including hardware devices like USB drives and paper wallets. Theoretically, if a hacker obtains either, he can take cryptocurrency from a user's digital wallet without the victim's knowledge [21].

Faulty Dates: Each node internally updates a number that denotes the Bitcoin blockchain's network time. It is based on the message when peers connect, which contains the median time of the node's peers. The network time counter, however, switches back to system time if the median time deviates from system time by more than 70 minutes. An attacker who connects as several peers and reports incorrect timestamps may affect the network time counter of a node [22].

Timestamp Hacking, or *Poison Pill*, would include slowing down the target's clock while speeding up the clock of most of the network mining resources. The difference between the nodes would be 140 minutes because the time value can only be distorted by a maximum of 70 minutes. Furthermore, nodes must have a reliable clock to connect to their peers and the network. If a node's time deviates by 12 seconds from synchronized UTC, the number of peers will decrease until it eventually has none left and is cut off from the network. However, the block's timestamp should not be used for vital parts of the smart contracts [22].

Vendor Supply Risk: Since the technology is still in its infancy, its use has had numerous restrictions and difficulties. Factors leading toward the adoption of blockchains include enterprise leaders' knowledge of blockchain. Even company leaders who recognize blockchain's potential are hesitant to dedicate time and resources to it due to the absence of industry-wide standards and practices. Also note that blockchains may not be supported by the traditional ERP systems used in many firms. So, a decisive move is needed to either set up in-house development or outsource the application development for its specific supply chain. As an organization must contract a third party whose services must be trusted with the organization's data, privacy leakage might be a serious worry among other concerns. Concerning technical challenges, when it comes to retrieving and committing records, blockchains are much slower. and use a lot more computational power.

The scalability of these resources has been a significant concern. Scalability means that the capacity of the system to respond and function even after the input size may be increased to meet user demand. A wider adoption of blockchains is being hampered by the isolation of blockchain as individual "silos" caused by the absence of interoperability standards. Resolving issues with collaboration and cross-chain interaction between public, private and consortium blockchains can soon open possibilities for a hyper-connected world [23].

Blockchain network risks

Network attacks refer to events that threaten blockchain networks by exploiting flaws in their communication protocols, design or implementation [24]. In 2013, hackers stole over 4,000 Bitcoins from *Input.Io* in two separate attacks in the same week. Input.Io stores cryptocurrency in digital wallets for its customers. The attack was made possible by a flaw in the cloud infrastructure. Furthermore, the attacker changed the *LinodeServer* password and compromised the client's e-mail accounts. Selfish mining attacks, block withholding attacks, Sybil attacks, transaction malleability attacks, and Eclipse attacks are considered network security risks [22].

Selfish Mining Attack: A dishonest miner (or miners) will not broadcast and share the legitimate solution to the network as part of a selfish mining attack. In this attack, the dishonest miners selectively release blocks or publish multiple blocks at once, forcing the other miners in the network to trash their blocks and lose money. The primary motivations of selfish mining would gain unfair rewards being more significant than their share of computer power spent and confuse and lead other honest miners to waste their resources in the wrong direction.

Block Withholding Attack (BWH): Blocks have been detected during block withholding attacks, and dishonest miners never broadcast a mined block to harm the pool's revenue. Attacks that use block withholding typically include infiltrating another pool [24].

Forth after Withholding Attack (FAW): FAW is the other type of BWH attack. The FAW attack has hybridized selfish mining and the BWH attack [24]. The attackers rewarded for a FAW attack have been equal to or higher than a BWH attacker's payoff, thus being four times more beneficial per pool than the BWH attack [24].

Transaction Malleability Attack: This type of attack has been classified as a double spending attack. However, in this case, the attacker becomes the transaction receiver, and the victim to be the sender. The attacker would cause the latter to create a transaction that transfers some value to a second deal controlled by the attacker. The attacker then modifies the identification hash while leaving the transaction unaffected. The new transaction gets transmitted to the community. If a modified version of the transaction gets later verified, the malleability attack may succeed. The victim's money has been doubled by the assailant. Custom implementation of Bitcoin was susceptible to this attack. For instance, MtGox, the once-largest Bitcoin exchange, shut down and filed for bankruptcy in 2014, claiming that attackers exploited malleability attacks to empty its accounts [22].

Sybil Attack: Distributed networks can be susceptible to the Sybil attacks. This attack can be particularly effective when peer nodes remain anonymous or operating under a pseudonym and participation remains open. The purpose of the attack is to isolate a user from trustworthy network nodes to set the stage for subsequent attacks [22]. The general type of attacks on peer-to-peer networks includes the creation and control of numerous false identities by a single malicious actor. This kind of attack can be employed in blockchain networks to isolate a target node from the rest of the trustworthy network, which in turn is utilized to launch other attacks [13]. For example, refusing to send transactions that were initiated by an honest node but were not validated by the network because their input may be exploited to commit double spending; or, planning a timing attack, which entails keeping track of communications coming from the victim and going to the attacker's ISP with the intention of getting around the low latency encryption and anonymization of the transmission [22].

Eclipse Attack: In an eclipse attack, all incoming and outgoing connections to the target would be monopolized to isolate the user from the rest of the network. This attack method gives the attacker a chance to tamper with the target's perception of the blockchain, making it use its computing resources inefficiently, or use the target's computing resources for malicious purposes. The only node that accepts incoming connections would be the attack target node [13].

Byzantine failure

A decentralized system's ability to tolerate Byzantine faults has been known as *Byzantine Fault Tolerance (BFT)*. A crash failure occurs when nodes simply stop acting in any way, whereas Byzantine failure occurs when nodes just don't do anything or exhibit arbitrary behavior. Crash failure is basically a subset of Byzantine failure. An imminent risk is that one or more rogue or untrustworthy actors will cause the environment to disband in any decentralized computing environment that uses a blockchain data structure. A server in a cluster will not function properly if several of its servers fail to send data to other servers on a consistent basis [25].

The working of PBFT (Figure 3.1) is as follows:

1. The transaction is requested by the client and is broadcast to the network. The leader will be chosen in a *Round Robin* fashion.
2. The state shifts to a pre-prepared state as the leader prepares a block proposal and broadcasts it to the network.
3. After the backup nodes start receiving the pre-prepare information and confirm the request, they broadcast messages throughout the network.
4. The node commits after receiving the same results from the $2m+1$ nodes.
5. The block will be added to the chain if $2m+1$ nodes broadcasts commit messages and then advance to the last committed state after committing a block to the chain.

To be reliable, the decentralized computing environment must be designed in such a way that can handle Byzantine failures [25].

Environmental perturbation

Assuming that prices and user acceptance continue to fluctuate, the amount of energy used by Bitcoin mining will probably change over time. The incentive to start mining also rises as the value of the block reward does making cryptocurrency mining a competitive activity [26]. In addition, computational cost/energy implies the overall expense related to transaction verification. Both the mining method and necessary bandwidth for block propagation are included. On a permissionless blockchain, the mining process becomes connected to the consensus mechanism. Bitcoin employs the PoW consensus technique, which necessitates specialized hardware for block mining. A public blockchain's implementation scale may be impacted by the specific high-end hardware's increased energy consumption, which overall results in higher computing costs [10]. Although most of the nations where cryptocurrencies are mined use fossil fuels as their main energy source, miners must look for the least

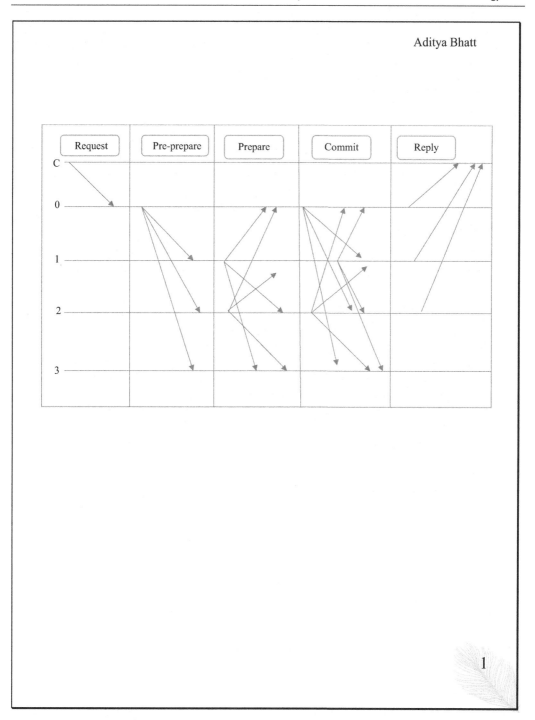

Figure 3.1 PBFT algorithm.

Source: [18] (Created with MS Word)

expensive energy sources to stay profitable. According to previous research, *Turkmenistan* produces the same amount of carbon dioxide every year as the Bitcoin network or around 73 million tons. According to data up until September 2022, Ethereum generated an estimated 35.4 million tons of carbon dioxide emissions while switching to proof of work, after which those emissions decreased to 0.01 million tons. According to a researcher at the University of Cambridge, most of the Bitcoin mining takes place in China and the United States. Due to the rapid obsolescence of mining equipment, cryptocurrency mining also produces a substantial amount of electronic waste. This became particularly valid for *Application-Specific Integrated Circuit* (ASIC) miners, which are specialized devices made for mining the most widely used cryptocurrencies. The Bitcoin network, according to researchers, produces over 38 thousand tons of electronic waste per year [26].

Legal-compliance risk

Legal compliance may typically be challenging to determine which government's rules and regulations apply to specific blockchain applications, because the nodes of a decentralized ledger can span numerous locations worldwide [4]. Transactions carried out by an organization run the risk of falling under the laws of every country where a node in the blockchain may locate, which might result in an overwhelming number of laws and regulations that could apply to transactions in a blockchain-based system [27]. The *US Securities and Exchange Commission* (SEC) has voiced worries that many initial coin offerings (ICOs) will be fraud attempts or attempts to raise money in violation of investor protection rules. Policymakers and regulators in other nations have attempted to explain the situation by stating that not all initial coin offerings would be subject to the same investor protection rules as an IPO. With the permissionless blockchain system, all users may have access to the data on a network and no one entity oversees maintaining the network's availability or security. Determining whether a party qualifies as a controller or processor when processing personal data is crucial because GDPR places different obligations on a controller (the party that chooses the purposes and methods of processing specific personal data) and processor (a party that processes personal data on behalf of a controller, such as an outsourced service provider). Based on their respective activities, it is essential to consider the degree to which various blockchain network players are controllers. As mentioned, a lack of rules and standards is one of the main problems in implementing blockchain for many firms. It is simpler to establish a legal framework and internal governance structure in a permissioned or private system that will specify the governing legislation that will apply to transactions. It would also be advantageous to consider some kind of mutually agreed-upon dispute resolution procedure in private systems [28].

CONCLUSIONS AND RECOMMENDATIONS

Blockchain technology has already emerged as one of the most promising technologies and research topics of the 2020s. As such, the purpose of this chapter was to provide a brief overview of the operational risks involved in blockchain technology, as well as the means by which they can be mitigated. It is the aim of this chapter to give readers a fundamental understanding of the operational risk associated with various domains in which blockchain technology is applied. A variety of topics were discussed, including the origin of operational risk, how it impacts various aspects of blockchain technology, and effective practices to mitigate those risks. Through interaction with software systems, web-based systems, cloud and other technologies blockchain systems can introduce security risks. Therefore,

It must be used in conjunction with an adequate security framework. The incident examined in this chapter highlights the need for important elements of the blockchain-specific framework, such as legal compliance, the necessity of smart contract code reviews and internal-external audits, automation of IR methods and checks, and appropriate use of cold storage techniques.

Recommended operational risk mitigation practices

This chapter examined operational risks in blockchain technology. Good practices cover the four major areas of risks in blockchain such as people risks, process risks, system risks, network risks, consensus risks and legal-compliance risks, which are as follows:

People Risk means on whom companies become overly reliant, including those with highly technical jobs and the creative minds behind design work.

Remove Dependency from Key Persons: It creates short-term and long-term risks for an organization. Determine the level of risk that reliance on those individuals poses to the organization. To solve key person issues, an organization must first determine what critical information certain employees may bring to the table that no other employee has. If a key person leaves the organization, the organization must devise a method a pass that information on. Varieties of training programs through institutions and organizations may fill the shortfall of talented

As the *Process Risk* description portrayed how software bugs are vulnerable to the entire blockchain, core software code security is necessary using the following procedures:

The Need for Periodic Smart Contract Audits and Improvements: For example, the two-round multi-signature technique has been found to have vulnerabilities in 2019, and *Drijvers et al.* presented *mBCJ* as a provably secure yet highly effective replacement. In 2020, *Sun et al.* presented *Counter-RAPTOR* to counteract and identify active routing threats, while Drijvers et al. presented *Pixel*, a pairing-based forward-secure multi-signature method. An additional step before the smart contract is deployed is an audit of the smart contract. In 2018, *Erays* (researcher) was presented to reserve engineering the smart contract into high-level pseudocode and manually examining various security contract features [14].

The Need to Monitor and Analyze Transaction Logs: The first generic framework to perform bytecode-level, logic-c-driven analysis on Ethereum transactions for attack detection in 2020 was *TxSpector*. Attacks such as Reentrancy, Unchecked Call, Timestamp Dependence, Misuse-of-Origin, Failed to Send, Mishandled Exception, Unsecured Balance and DoS were detected [14].

Recommended system control risk mitigation practices

At the core of the blockchain, system control risks mainly tie up with scalability issues, interoperability with various applications, major security failures and controls over wallet risks.

To address the blockchain scalability challenge, consider the following tools and techniques:

Segwit is the method where block size will be kept constant when adding more transactions. To successfully create more room for new transactions, this strategy seeks to separate the signature information from the transaction and store it elsewhere.

This method keeps the validating portion and the transaction's actual data apart. Given that the digital signature makes up about 70% of the transaction, it is necessary to keep the signature in the witness data structure, which is separate from the transaction, to address the malleability of the transaction.

Next, in order to optimize storage in commercial databases, *Sharding* has historically been recommended in the database sector. It is considered one of the most efficient approaches. The nodes are divided into numerous pieces known as shards when using the sharding approach. Each shard possesses a small portion of a transaction. As a result, the transaction is handled concurrently. The transaction is processed in parallel, which enhances the authentication system and ultimately increases the throughput of the entire blockchain network.

The interoperability of blockchain technology with various sectors and associated applications depends on several factors, such as organizational readiness, technical expertise and digital infrastructure. Social media platforms based on crowdsourcing or the Internet of Things (IoT) and blockchain can help to increase industry resilience and agility. The security of the information management system, consensus mechanisms for dynamic data storage, transparency, data protection and dependability may all be improved with the efficient integration and deployment of blockchain with IoT and Radio Frequency Identification (RFID) [29, 30].

Table 3.3 represents information about various network risks and their role and good practices that address the risks.

Table 3.3 Good practices for network risks mitigation

Security threat	Attack vector	Affected abstract layer	Good practices
Network Threats	Block Withholding and Forth-After Withholding Attacks	Network, Consensus	Only trusted miners should be permitted to register by the pool management. Also, the pool should be shut down if its computational effort results in lower revenue than anticipated.
	Transaction Malleability Attack	Consensus, Data Model	To address this problem and apply the new transaction validity rule that takes into account several metrics for transaction verification, BIP (Bitcoin Improvement Proposal 62) was planned to be implemented as a soft fork.
	Sybil Attack	Network	PoW prevents misleading information from being fed to the victim in a successful Sybil attack.
	Eclipse Attack	Network	Diversify incoming connections from the same IP address. Set ADDR message acceptance limits such as large unsolicited ADDR messages (with >10 addresses) from incoming peers may not be accepted by a node.
	Selfish Mining Attack	Network, Consensus	To reduce likelihood of the selfish mining attack, Bitcoin improvement proposals suggested randomly allocating miners to different branches of pools. Freshness Preferred (FP) to the timestamp that already existed in each block header.

Source: [20][24]

Recommended legal and compliance good practices

Regulators may soon be forced to abandon specific and prescriptive rules in favor of generally expressed principles that cover the maximum standards at which the industry functions. The regulators should collaborate closely with stakeholders to ensure that their views on legal business practices are fair, open and transparent. This would make it possible for legislation to be adaptable enough to cover the vast array of systems that blockchain technologies enable.

REFERENCES

1. Patel, S., Sahoo, A., Mohanta, B. K., Panda, S. S., & Jena, D. (2019, March). DAuth: A decentralized web authentication system using Ethereum-based blockchain. In *2019 International Conference on Vision Towards Emerging Trends in Communication and Networking (ViTECoN)* (pp. 1–5). IEEE. https://ieeexplore.ieee.org/stamp/stamp.jsp?tp=&arnumber=8899393
2. Odeh, A., Keshta, I., & Al-Haija, Q. A. (2022). Analysis of Blockchain in the Healthcare Sector: Application and Issues. *Symmetry*, 14(9), 1760. From, https://www.mdpi.com/2073-8994/14/9/1760
3. *Deloitte: Blockchain Risk Management Risk Function must play an active role in shaping blockchain strategy.* From https://www2.deloitte.com/content/dam/Deloitte/us/Documents/financial-services/us-fsi-blockchain-risk-management.pdf
4. Baiod, W., Light, J., & Mahanti, A. (2021). Blockchain technology and its applications across multiple domains: a survey. *Journal of International Technology and Information Management*, 29(4), 78–119. From, https://scholarworks.lib.csusb.edu/cgi/viewcontent.cgi?article=1482&context=jitim
5. Office of the Superintendent of Financial Institutions. (2016, June). *Operational Risk Management.* https://www.osfi-bsif.gc.ca/Eng/fi-if/rg-ro/gdn-ort/gl-ld/Pages/e21.aspx
6. Robert Sheldon. (2021). *6 must-have blockchain developer skills from,* https://www.techtarget.com/whatis/feature/6-must-have-blockchain-developer-skills
7. Uriel Barillas. (2022). *The Risk of Key Person Dependency for Information from,* https://www.forbes.com/sites/forbestechcouncil/2021/12/28/the-risk-of-key-person-dependency-for-information/?sh=147c464b1241
8. Flensburg, Schleswig-Holstein. (2022, September 21). *The Blockchain Talent Shortage Crisis – How This Company Is Working To Fill The Blockchain Developer Void.* https://newsdirect.com/news/the-blockchain-talent-shortage-crisis-how-this-company-is-working-to-fill-the-blockchain-developer-void-255710497
9. Pawan Nahar. (2022, May 03). *Despite the rising learning platform, the crypto industry facing a shortage of the right talent.* https://economictimes.indiatimes.com/markets/cryptocurrency/despite-rising-learning-platforms-crypto-industry-facing-shortage-of-right-talent/articleshow/91280790.cms
10. Khan, D., Jung, L. T., & Hashmani, M. A. (2021). Systematic literature review of challenges in blockchain scalability. *Applied Sciences*, 11(20), 9372. From, https://www.mdpi.com/2076-3417/11/20/9372
11. Zamani, E., He, Y., & Phillips, M. (2020). On the security risks of the blockchain. *Journal of Computer Information Systems*, 60(6), 495–506. https://www.tandfonline.com/doi/full/10.1080/08874417.2018.1538709
12. Nguyen, G. T., & Kim, K. (2018). A survey about consensus algorithms used in blockchain. *Journal of Information processing systems*, 14(1), 101–128. https://scholar.google.com/scholar_lookup?title=A%20Survey%20about%20Consensus%20Algorithms%20Used%20in%20Blockchain&author=Giang-Truong%20Nguyen&publication_year=2018
13. Dasgupta, D., Shrein, J. M., & Gupta, K. D. (2019). A survey of blockchain from a security perspective. *Journal of Banking and Financial Technology*, 3(1), 1–17. From, https://link.springer.com/content/pdf/10.1007/s42786-018-00002-6.pdf

14. Guo, H., & Yu, X. (2022). A Survey on Blockchain Technology and Its Security. *Blockchain: Research and Applications*, 3(2), 100067. From, https://reader.elsevier.com/reader/sd/pii/S2096720922000070?token=273E6A9583359C22BFCBB6C19994EF95F5DE080834AEF57D6257B644E5C3425C3D059FE0EADAD936C6F3B61CDEB1EC83&originRegion=us-east-1&originCreation=20221029192431

15. Kenny L. (2019). *The Blockchain Scalability Problem and the race for VISA-Like Transaction speed.* https://towardsdatascience.com/the-blockchain-scalability-problem-the-race-for-visa-like-transaction-speed-5cce48f9d44

16. Casino, F., Dasaklis, T. K., & Patsakis, C. (2019). A systematic literature review of blockchain-based applications: Current status, classification and open issues. *Telematics and informatics, 36,* 55–81. https://www.sciencedirect.com/science/article/pii/S0736585318306324

17. Mackay, B. *Evaluation of Security in Hardware and Software Cryptocurrency Wallets.* https://www.researchgate.net/profile/Brian-Mackay/publication/338582744_Testing_design_and_methodology/links/5f6a456f92851c14bc8e1e8a/Testing-design-and-methodology.pdf

18. H. Kohad, S. Kumar and A. Ambhaikar, "Consensus Algorithms in Blockchain Technology," *2021 6th International Conference on Signal Processing, Computing and Control (ISPCC)*, 2021, pp. 159–164, doi: 10.1109/ISPCC53510.2021.9609412. https://ieeexplore.ieee.org/document/9609412

19. Samy, H., Tammam, A., Fahmy, A., & Hasan, B. (2021). Enhancing the performance of the blockchain consensus algorithm using multithreading technology. *Ain Shams Engineering Journal, 12*(3), 2709–2716. https://reader.elsevier.com/reader/sd/pii/S2090447921000630?token=D48A5422A974DE8DDC816C4CC1FC6C5FEA807155950D61E537DBF271683BD88A005DA9699CAAC5A391DE3CA2C09E7A8C&originRegion=us-east-1&originCreation=20221031181756

20. Chaudhry, N., & Yousaf, M. M. (2018). *Consensus algorithms in Blockchain: Comparative Analysis, challenges and opportunities.* IEEE Xplore. Retrieved November 2, 2022, from https://ieeexplore.ieee.org/stamp/stamp.jsp?tp=&arnumber=8632190

21. Gupta, S. (2022, September 20). *How blockchain can be hacked: 5 ways to hack the blockchain.* Tech Research Online. Retrieved November 3, 2022, from https://techresearchonline.com/blog/blockchain-can-be-hacked/

22. Morganti, G., Schiavone, E., & Bondavalli, A. (2018, October). Risk assessment of blockchain technology. In *2018 Eighth Latin-American Symposium on Dependable Computing (LADC)* (pp. 87–96). IEEE. https://ieeexplore.ieee.org/stamp/stamp.jsp?tp=&arnumber=8671548

23. Jabbar, S., Lloyd, H., Hammoudeh, M., Adebisi, B., & Raza, U. (2021). Blockchain-enabled supply chain: analysis, challenges, and future directions. *Multimedia systems, 27*(4), 787–806. https://link.springer.com/article/10.1007/s00530-020-00687-0

24. Mosakheil, J. H. (2018). *Security threats classification in blockchains.* https://repository.stcloudstate.edu/cgi/viewcontent.cgi?article=1093&context=msia_etds

25. Prusty, N. (2018, September). *Blockchain for Enterprise.* O'Reilly Online Learning. Retrieved November 2, 2022, from https://learning.oreilly.com/library/view/blockchain-for-enterprise/9781788479745/78fb3794-8a21-4303-b10b-4715d7cc5fd8.xhtml

26. Nathan R. (2022). *What's the Environmental Impact of Cryptocurrency?* https://www.investopedia.com/tech/whats-environmental-impact-cryptocurrency/

27. Abughosh, S. (2019, January 17). *Managing the risks of Blockchain in the financial industry.* LinkedIn. Retrieved November 3, 2022, from https://www.linkedin.com/pulse/managing-risks-blockchain-financial-industry-suha-abughosh/

28. Salmon, J., & Myers, G. (2019). *Blockchain and associated legal issues for emerging markets.* https://openknowledge.worldbank.org/bitstream/handle/10986/31202/133877-EMCompass-Note-63-Blockchain-and-Legal-Issues-in-Emerging-Markets.pdf?sequence=1&isAllowed=y

29. Dutta, P., Choi, T. M., Somani, S., & Butala, R. (2020). Blockchain technology in supply chain operations: Applications, challenges and research opportunities. *Transportation research part e: Logistics and transportation review, 142,* 102067. https://www.sciencedirect.com/science/article/pii/S1366554520307183

30. Zheng, Z., Xie, S., Dai, H., Chen, X., & Wang, H. (2017, June). An overview of blockchain technology: Architecture, consensus, and future trends. In *2017 IEEE international congress on big data (BigData congress)* (pp. 557–564). IEEE. https://ieeexplore.ieee.org/abstract/document/8029379?casa_token=kAwqgpEiD7sAAAAA:MtYM3XPp9cYD90LSf5KRt1ddup4cs0MtyhzXOeebCBZmRiMbe-VYsnQTO5vvXa4eO3brvpQh7Bw

Chapter 4

Privacy considerations in blockchain technology

Adebimpe Onamade and Ebubechi Okpaegbe

BACKGROUND

In the past, people used physical currency like stones or coins to exchange goods and services. However, on a small island called *Yap* in the South Pacific around 1000 BC, the Yapese people used a unique form of currency called *RaiStones*. These stones were rather large, standing at 12 feet (3–4 meters) tall and weighing over 8,000 pounds (3,500+ kg). To keep track of the ownership and exchange of these stones, the Yapese people used a ledger. The ledger recorded what was exchanged and who exchanged it, allowing for the ownership of the stones to be tracked without the need for physical exchange [1].

In a similar manner, in modern times, blockchain technology allows for the decentralized ledger to record all transactions, making it accessible to anyone. However, while the transparency of the ledger provides trust in the network, it also poses a risk to the privacy of consumer data. The ability for anyone to view every transaction on the blockchain can quickly become a tool for tracking people and their spending habits. This can lead to data breaches, which can be exploited by police and/or criminals to track down individuals through their digital assets [2]. To ensure the appropriate protections for data subjects, entities that leverage blockchain technology for executing transactions and storing information must consider how their information may be used or disclosed. While the transparency of the ledger is a valuable aspect of the blockchain, measures must be taken to safeguard the privacy of consumer data [3].

Privacy concerns

Data is fast becoming a new asset class that may surpass previous asset classes. Internet giants such as *Amazon*, *Google*, *Apple* and *Facebook*, now known as powerful *digital conglomerates*, have gained control over the vast amounts of data created by citizens and organizations, most often stored in central databases. While these companies offer value to consumers, this has led to data being considered a valuable asset that may surpass other assets. Furthermore, this has caused traditional notions of privacy and individual autonomy to be questioned [4]. Due to privacy concerns, 48% of individuals worldwide have stopped purchasing or using a company's products or services. Invariably organizations have generated billions in revenue from business models based on harvesting troves of personal data obtained by using a variety of tactics such as cajoling, luring and trickery. The *2021 Cisco Data Privacy Benchmark Study*, which carried out a survey of 4,700 security professionals in major industries from 25 countries stated that privacy budgets doubled in 2020. However, 87% of people worldwide are concerned about the privacy measures in the tools needed to work and connect remotely. During the COVID pandemic, 36% of customers were unwilling to give up their privacy for

DOI: 10.1201/9781003462033-5

safety. Australians also take extra precautions to protect their privacy by deleting an application, blocking access to their information or switching to a different service provider. The top factor for Australian customers when selecting digital services is privacy before dependability, convenience and affordability. As such, there is strong consensus that effective privacy management is a good business practice. Strong market forces will continue to push businesses of all sizes to prioritize privacy, promote transparency and build trust [3, 5].

Understanding data privacy

Data privacy or information privacy is a branch of data protection concerned with properly handling sensitive data, most notably personal data and other confidential data to meet regulatory requirements while protecting the data's confidentiality and immutability [6]. Privacy is the cornerstone of a free society, allowing individuals to choose which components of their identity they wish to share and with whom while maintaining ownership of their data [4]. How organizations seek consent to hold personal data, comply with privacy regulations and manage the collected data is critical to creating confidence with customers who demand privacy as a basic human right. Data privacy is also essential for business asset management and regulatory compliance. Business asset management involves understanding how people live in a data-driven economy. Organizations recognize data as a valuable asset and emphasize collecting, sharing and analyzing data about their clients and users, primarily through social media. Regulatory compliance ensures that personal data is managed correctly to meet legal obligations on data collection, storage and processing. Noncompliance can lead to hefty fines, which can cause significant financial losses and erode customer trust if a cyberattack occurs [6]. In short, privacy has become an essential component for successful and innovative enterprises and brands in the modern age, as enterprises use the new age of privacy as an opportunity to develop unique products and services, innovate and attain customer trust. It also has been suggested that privacy can be considered the equivalent of product safety in the digital age. This notion appears to have been reflected in the results of a study that found firms investing in privacy programs experienced an ROI of up to double their expenditure, offering a distinct competitive advantage [5].

Personal Data refers to any personal information relating to an individual that can be identified or is identifiable including individual details like name, number, address, telephone, credit card number, IP address or a cookie [7]. According to the General Data Protection Regulation (GDPR), data is considered *personal data* when an individual can be traced directly or indirectly by online identifiers such as their name, an identity number, IP addresses or location data; or, include information particular to that natural person's physical, physiological, genetic, mental, economic, cultural or social identity. Every engagement people have with organizations involves exchanging personal information; and, more than one of these pieces of information may be needed to identify a person when combined. Personal information used to identify a specific individual and constitute personal data is called *personally identifiable information* (PII). However, when data is anonymized, it ceases to be a PII, and an individual is no longer identifiable. However, for data to be genuinely anonymized, anonymization must be irreversible [8].

Data Subject is defined as the living persons to whom the data relates, but the GDPR extends beyond this to include any duty of confidence in place prior to the death of any individual [7]. Furthermore, a data subject can be an identified or identifiable natural person with any information that can be traced to them directly or indirectly [8]. *The General Data Protection Regulation* AKA Regulation (EU) 2016/679 gives data subjects control over their personal data, including how it is collected and stored and requires those who collect and store such data to agree to hand over, correct and delete that data upon request [9]. In the event of a data breach, GDPR mandates an assessment to determine whether there is a possible danger to the data subjects impacted, which must be completed within 72 hours and reported if there is a threat to affected people's rights and freedom [7].

The GDPR provides individuals, referred to as *data subjects* with control over their personal data, including how it is collected and stored, and requires those who collect and store such data to agree to hand it over, correct and delete it upon request. As mentioned, the GDPR extends beyond this to include any duty of confidence in place prior to the death of any individual. Furthermore, a data subject can be an identified or identifiable natural person with any information that can be traced to them directly or indirectly.

Data Controller is either a natural or legal person, an agency, a public authority or any other body that alone, or in collaboration with others, determines the goals and means of processing personal data [8]. A business or organization is referred to be a data controller when it collects data directly from the user [10]. The data controller is responsible for overseeing the management of personal data in order to adhere to GDPR [7]. Data controllers make critical decisions and have the final say and control over the reasons and purposes for data gathering and the means and methods of data processing [8]. The development of blockchain technology has made it possible for different parties to collaborate on a single blockchain. Article 24 of GDPR states that a data controller must consider the purpose, nature, context and scope of any data processing activity, the possibility of any serious threat to any natural person's liberties and rights, and implement suitable organizational, technological and security measures to demonstrate that data processing activities were carried out in conformity with GDPR. Data controllers are expected to carry out review and update measures where necessary, except for exempted data [8] [11].

Data Processor is any legal or natural person, organization, governmental authority or other body that processes personal data on behalf of and under the authority of a data controller [8]. The data processor is the person in charge of processing personal information by obtaining, recording, adapting, disclosing, using and ensuring data availability based on the instruction of the data controller [7]. Article 28 of GDPR states that if any data processing activities are carried out based on a controller's instruction, the data processor must implement appropriate organizational and technical measures to meet the guidelines set out by the GDPR. Data processors are responsible for ensuring that data subjects' rights are safeguarded. Thus, they should have their security measures in place [8]. Generally, the data processor participates in the more technical components of the procedure, while the data controller is responsible for interpreting and making the major decisions [7].

Anonymity and Pseudonymity are blockchain concepts used interchangeably but have different meanings. Anonymity means the technological inability to link actions performed with the person who performed the actions. This implies that because the person is unknown, their actions cannot be traced to them. Although blockchains enable actions to be seen based on decentralized technology, anonymity separates the person's identity from the actions performed. Anonymity means one's action cannot be linked to a face or name. At the same time, pseudonymity refers to digital activity taken on behalf of a fictional person whose name is typically unrelated to the natural person [12]. Concerning blockchain security, pseudonymity means that although the identity of the individual performing transactions is unknown, all their transactions may be traced to the same pseudonymous identity.

On the other hand, anonymity means that no transaction or activity on the blockchain or on exchanges can be attributed to a single user, pseudonymous or otherwise [13]. Pseudonymity refers to a state of disguised identity In Bitcoin. Addresses on a blockchain are hashes of public keys of a node (user) in the network. Users can interact with the system by using their public key hash as their pseudo-identity without revealing their real names. Thus, a user's address can be viewed as a pseudo-identity.

In addition, users can generate as many vital pairs (multiple addresses) as they want in a similar way as a person who can create multiple bank accounts. Although pseudonymity can achieve a weak form of anonymity utilizing the public keys, there are still risks of revealing the identity information of users [14]. As such, anonymity is a significantly stronger kind of privacy than pseudonymity. In any case, all parties will have to make compromises to guarantee a secure and safe cryptocurrency and blockchain security [13]. According to some blockchain advocates, using public–private key encryption maintains anonymity and privacy; however, this approach to managing personal information is unacceptable under most privacy laws [15]. *Territorial and Cross-Border Data Transfer* refers to data privacy rules applied to an individual's location where personal data are being processed. Irrespective of the jurisdiction where personal data are being processed, specific safeguards and adequate protection must be ensured at all levels. More sensitive blockchain projects that handle personal data may try to limit users by jurisdiction. However, validating online locations accurately can be challenging. The decentralized nature of blockchain technology not only affects the implementation of various jurisdictions' laws but creates tensions with those who impede cross-border data transfers [15].

Legitimate Bases for Processing Personal Data refer to reasonable justifications for processing personal data, as some data protection and privacy regulations require. For example, in the United States, Federal sector-specific rules like the Gramm-Leach-Bliley Act of 1999 (*GLBA*) and *the Health Insurance Portability and Accountability Act of 1996 (HIPAA)*, as well as various state laws, prohibit the use of some personal data without the consent of individuals. However, there may be exceptions, such as HIPAA, where there are basis and approved uses for treatment, payment and healthcare operations [15].

The unlinkability of transactions in a system implies that there is an inability to accurately ascertain the connection between two observations or entities in the system. It is imperative that users are ensured that any transactions related to them cannot be linked. If a user's transaction is related to another, it is possible to deduce additional information about the user such as their account balance and the type and frequency of their transactions. Furthermore, anyone can link a user's transaction to other transactions involving their account through a basic statistical analysis of Bitcoin addresses. Therefore, the complete anonymity of a user can only be protected by ensuring both pseudonymity and unlinkability. Unlinkability makes it hard to launch de-anonymization inference attacks, which link a user's transactions together to uncover the user's true identity in the presence of background knowledge [14].

Deanonymization enforces anonymity in transactions. For example, Bitcoin allows users to generate multiple Bitcoin addresses, and it only stores the mapping information of a user to her Bitcoin addresses on the user's device. One needs to construct a one-to-many mapping between the user and its associated addresses to deanonymize [14]. While users may employ many identities (or pseudonyms) to enhance their privacy, an increasing body of research shows that identities, at times, can be deanonymized on the Bitcoin blockchain by analyzing the structure of the transaction graph and the values and dates of transactions [23].

Privacy frameworks, laws, and regulations

Over the past several years, more than 60 governments worldwide have implemented or proposed new privacy and data protection legislation (including Argentina, Australia, Brazil, Egypt, India, Indonesia, the United States, Singapore, Japan, Mexico, Nigeria, Panama, Kenya and Thailand). This trend is expected to continue, with *Gartner* predicting that by 2023, 65% of the world's population will be covered by current privacy standards. Prior to the implementation of the GDPR in May 2018, organizations often paid relatively little attention to privacy issues, and there were only a minimal number of official privacy officers around the world. However, it was predicted that 500,000 organizations will have to depend on the expertise of a privacy officer by 2019 [5].

The various privacy laws and regulations that have been established in some countries are discussed in this next section:

The GDPR was implemented in April 2016 to ensure uniform data privacy standards across the European Union, granting individuals greater security and rights than the previously available laws. As mentioned, GDPR establishes guidelines for how companies and other organizations manage information about the people who interact with them [7]. The GDPR specifies how organizations and enterprises should manage information relating to the persons with whom they engage, which made it simpler for European Union individuals to realize their rights around personal information and its usage. As a result, personal information is more susceptible than ever [8]. From the perspective of EU citizens, the intention of GDPR is to make it more straightforward to comprehend how data will be employed prior to collection and to be able to submit a grievance no matter where the data is stored. Note that the objective of GDPR is to guard data that belongs to European citizens and inhabitants. As such, its regulations are applicable to organizations that handle this type of data, irrespective of whether the individuals are located in the European Union. Non-European entities must adhere to GDPR when selling merchandise or services.

The GDPR seeks to protect the data of European Union customers and reduce the risk of mishandling or misprocessing of personal data on the web by defining new concepts of personal data, types of consent, accountability criteria and roles involved in data decision-making, interpretation and processing. With this, GDPR aims to minimize the severity and frequency of data breaches that occur within businesses. The GDPR is a legal framework established around seven principles that provide directives to those who must comply with its requirements and expectations for how European Union (EU) citizens' data should be treated. These principles are *lawfulness, fairness, and transparency; purpose limitation; data minimization; accuracy; storage limitation; integrity and confidentiality; and accountability.* Additionally, GDPR guarantees individuals certain rights when it comes to the management of their data, such as the right to be informed, the right to access, the right to rectification, the right to erasure, the right to restrict processing, the right to data portability, the right to object and rights concerning automated decision-making and profiling. Organizations with fewer than 250 employees are exempt from certain record-keeping requirements but must still ensure the protection and security of EU citizens' data. In the event of a data breach, GDPR mandates the assessment of a potential risk to the affected individuals' rights and freedoms within 72 hours and the need to report any such risk [7].

The Personal Information Protection and Electronic Documents Act (PIPEDA) applies to private-sector entities in Canada that collect, use or disclose personal information during a commercial activity. Commercial activity is defined by the legislation as any specific transaction, act or conduct; or, any regular course of conduct that is of a commercial nature, including the selling, trading or leasing of donors, membership or other fundraising lists. PIPEDA applies to all enterprises that operate in Canada and covers personal information that crosses provincial or national boundaries during commercial activity, regardless of where they are located (including provinces with substantially similar legislation). Personal information is defined by PIPEDA as any factual or subjective information, recorded or not, regarding an identified individual under PIPEDA. This contains any type of information, such as age, name, ID numbers, income, ethnic origin, or blood type; employee files, credit records, loan records, medical data, existence of a dispute between a customer and a merchant, intents (e.g., to acquire goods or services, or change jobs) [16]. PIPEDA stipulates ten fair information principles that must be followed by businesses in contributing to building trust in business and the digital economy. *Accountability* requires businesses to appoint a responsible individual to ensure compliance with PIPEDA, protect personal information, develop and implement related policies and procedures, as well as handling any transfers of personal information to third parties. *Consent* requires meaningful consent from the data owner in the collection, usage and disclosure of personal information. Data owners have the right to withdraw consent and be informed of the implications of doing so. Limiting collection requires businesses to only collect personal information for legitimate purposes in a fair and lawful manner. *Limiting use, disclosure and retention* requires businesses to only use or disclose personal information for the purpose for which it was obtained, unless disclosure is consented to by the data owner or required by law. *Accuracy* requires businesses to reduce the probability of using inaccurate information when making decisions or releasing. *Safeguards* ensure personal information is safeguarded in a manner proportionate to its sensitivity and protect all personal information from loss, theft or unauthorized access, disclosure, copying, use or alteration (regardless of how it is kept). *Openness* entails organization privacy policies being detailed, clear, easy to understand and available when needed. *Individual Access* empowers individuals to access their personal information and challenges the accuracy, completeness and amendment of information held by the organization about them. *Challenging compliance* refers to the rights of an individual to challenge an organization's compliance with the principles of fair information and be directed to the person in the organization who is responsible for PIPEDA compliance [16].

The Health Insurance Portability and Accountability Act of 1996 is a federal law that requires the establishment of national standards to protect patients' health information from being revealed without their permission. To carry out this mandate, the *US Department of Health and Human Services* (HHS) established the HIPAA Privacy Rule and the HIPAA Security Rule. The Privacy Rule outlines guidelines for how entities subject to HIPAA are allowed to use and disclose individuals' health information (also known as protected health information or PHI). Additionally, it grants individuals the right to control how their health information is used; and, it sets standards to ensure the security of health information to protect the public's health and well-being. The Privacy Rule allows for vital information use while respecting the privacy of those seeking care and recovery. HIPAA applies to several entities, including healthcare providers, health plans and healthcare clearinghouses; however, a group health plan with fewer than 50 participants is considered exempt. Furthermore, a covered entity must obtain individual authorization to use and disclose PHI except in certain cases, such as public health activities, research under specific conditions or when dealing with an abuse victim. HIPAA compliance requires safeguarding the confidentiality, integrity and availability of all e-PHI protecting against anticipated illegal uses or disclosures and detecting and protecting against anticipated threats to the security of the information. The violation of HIPAA may result in civil, monetary or criminal penalties [17].

New Zealand's New Privacy Bill implemented the recommendations from the Law Commission issued in 2011, which address a wide range of issues, including the law's extra-territoriality, new mandatory data breach notification requirements, changes to the Office of the Privacy Commissioner's key enforcement tool, compliance notices, data subject access requests, restrictions on cross-border transfers of personal information and the general enforcement regime. Personal data is any information related to an identifiable, living individual. This can include names, email addresses, biometric data, as well as IP addresses, unique identifiers, search and browser histories, data related to the device, operating system, updates, location data, purchase and online shopping history, settings and website preferences and behavioral data such as scrolling and hovering mouse cursor. The GDPR sets out certain obligations for organizations to comply with, such as informing individuals of data collected, respecting individual rights, notifying authorities of data breaches, as well as being cautious of cross-border data transfers. New Zealand's Privacy Bill attempts to regulate the flow of personal information between countries and organizations, but it may not be as powerful as the GDPR or other privacy laws, as it does not provide users with the ability to opt in or out of data tracking, does not give individuals the right to be forgotten, and does not empower a privacy commissioner to impose sanctions for privacy breaches. This privacy bill also introduces criminal offenses for those who commit privacy offenses, with violators liable to a fine of up to $10,000 [18].

Lei Geral de Proteção de Dados (LGPD), or Brazilian General Data Protection Law, is the first comprehensive data protection regulation that affects the use of personal data in Brazil, both online and offline in the private and public sectors. It was passed into law on September 18, 2020, and backdated to take effect on August 16, 2020. LGPD applies to data processed within Brazil, its territory, as well as data belonging to individuals within Brazil irrespective of the location of the data processor. *The LGPD* introduces a set of principles and options for data processing. The principles are *purpose limitation, adequacy, necessity, quality of data, transparency, security, non-discrimination, accuracy, prevention, and the principle of accountability.* This legislation requires permission from the data subject for the collection and use of data, utilizing an opt-in model. Additionally, LGPD provides data subjects with a range of rights, such as the right of access, rectification, cancellation or exclusion; the right to object to processing; the right to revoke consent; the right to information and explanation about data use and the right to data portability. LGPD also mandates the appointment of a *Data Protection Officer (DPO)* by entities that process data, and an officer to facilitate communication between lawmakers and data subjects. LGPD requires privacy and data protection rights and principles to be considered during the conception, design and execution of products and services. Its penalty includes notices or fines of up to 2% of the company's earnings in Brazil in the previous fiscal year with a cumulative penalty of 50 million reals (about $13,305,657) per infringement [18].

The Personal Data Protection Act (PDPA) was established on January 2, 2013, as Personal Data Protection Council (PDPC), and several amendments have been passed to PDPA, which took effect on November 2, 2020, and February 1, 2021, respectively [2].

The Personal Data Protection Act of Singapore, which came into effect in 2014, includes several requirements that are intended to regulate the collection, use, disclosure and storage of personal data within the nation. It also calls for the establishment of the *Do Not Call Registry,* which allows individuals to opt out of receiving unsolicited telemarketing communications. The PDPA applies to private entities that are engaged in handling personal information, with the exception of usage for personal purposes as well as employees in the course of their employment and public and government agencies. It defines personal data as a wide-ranging concept that applies to any data, whether true or false, regarding an individual who can be identified from the data, or from that data along with additional details to which the organization has or is expected to have access. This includes but is not limited to names, addresses, email addresses, telephone numbers, IP addresses, cookie identifiers, unique IDs, search history, browser history, device data, location data, and information concerning age, gender, race, health, sexual orientation, appearance, political and religious convictions. Under the PDPA in Singapore, there are eight key obligations that must be adhered to by organizations. Penalties for noncompliance are 10% of the annual turnover of an organization with an annual turnover that exceeds $10 million or $1 million, whichever is higher [18].

The China Personal Information Protection Law (PIPL) is the first China comprehensive law that is designed to protect and regulate personal information. The law was passed on August 20, 2021, by the People's Republic of China (PRC) during the 30th meeting of the *Standing Committee of the 13th National People's Congress (NPC).* However, PIPL went into effect on November 1, 2021. PIPL is extraterritorial in scope and applies to entities that provide products or services, analyze and evaluate the activities of domestic natural persons and other circumstances as provided by laws and administrative regulations [18]. PIPL states that personal information must be processed in line with the principles of legality, legitimacy, necessity and good faith, and must not be obtained via deception, fraud, force or other methods [19]. It is applicable to both living and relatives of deceased data subjects and includes several key obligations. These include *the right to be informed about and decide on the processing of their personal information, as well as the right to object to or limit such processing; the right to have access and copy their personal information; the right to request that their data be corrected by the data processor; the right to erase information under one of the conditions specified in Art. 47; and the right to require that data processors explain their processing rules in response to data subjects' requests.* In addition, PIPL requires consent for the lawful processing of personal information on the following grounds: when fulfilling a contract or for employment purposes; to complete a legal obligation; during emergencies in protecting life, health, and properties of individuals; under reasonable media scope or relations actions under public opinion and interest; and in circumstances provided by law. An individual should be informed about data processing activities, including data processors or controllers' names and contact, purpose and methods of processing, type of personal information collected or processed, data retention period, method of exercising data rights and changes to key data processing elements [18]. Passing a security assessment approved by the *Cyberspace Administration of China* (CAC), obtaining personal information protection certification from a professional institution, and adhering to standard contractual clauses defined by CAC and other provisions of other laws and regulations are conditions to be filled for cross-border transfer of information. Penalties for noncompliance to the law can go up to 5% of the organization's previous year's revenue or 50 million yuan (around $7.7 million) and personnel directly responsible are subject to fines of between 100k and 1 million yuan [18].

Thailand's Personal Data Protection Act 2019 took effect on June 1, 2012; and has an extraterritorial scope that applies to the collection, usage and disclosure of personal data in Thailand or Thai residents irrespective of whether the organization was registered under Thai law or present in Thailand. It defined personal data as any information pertaining to a person that may be used to identify that person, whether directly or indirectly, but does not include information on deceased people and the law does not list many specific examples of personal data. In addition, it requires that personal data transfer should only be done to countries with high privacy standards and provides another basis for data transfer, which includes compliance with legal and or compliance obligations, contracts, vital interests and when carrying out the crucial task of public interest. The PDPA outlines exemptions for personal data collected by government agencies in relation to national security, money laundering and cybersecurity; as well as data collected by members of parliament, Judiciary and credit bureaus for private purposes. The key obligations under the PDPA include obtaining *user consent, storing records of consent for five years, preventing unauthorized access, and ensuring the highest level of security and privacy standards.* The PDPA requires that users' rights be respected, such as being informed of the purpose of data collection, withdrawing consent, accessing and requesting data collected from individuals, raising objections to data collection and use, and requesting for data to be erased, destroyed or anonymized. Violations of the rule may result in a fine of THB 5 million ($159,591) or up to 4% of global turnover and criminal sanctions, including imprisonment for up to one year [18].

Switzerland's new Federal Act on Data Protection (nFADP) became effective in Fall of 2023 and provides new measures on consent, processing records, data breaches and data protection impact assessment, so that companies that are currently complying with GDPR have an edge in preparing for nFADP. *Swiss FADP* applies to individuals and businesses processing personal data and offering products and services that are incorporated in Switzerland. Its focus is on protecting the personal data of natural individuals, and not legal entities. Swiss nFADP defined personal data as all and any information that can be used to identify a person, and it includes full names, a picture showing the person's face, email address, telephone number, social security number and other sensitive information; including religion, ideological data, health data, racial origin, genetic and biometric data. The Swiss nFADP imposes certain obligations on data controllers. These obligations include the disclosure of data processing activities to data subjects, which covers the purpose of data processing, information on who will receive their data, and the contact details of the responsible officer; the maintenance of records of processing activities; the appropriate collection of user consent; the implementation of processes to address data breaches; the compliance with requirements for cross-border data transfers; and the conduct of data protection impact assessments, if and when required. Organizations that pass a specific threshold are not obligated or mandated to appoint a data privacy officer. However, they are encouraged to have a data protection advisor. The Swiss nFADP penalizes the person and not the organization that violates the obligations spelled out in the privacy act and imposes a fine of up to 250,000 CHF [18].

Privacy laws' applicability to blockchain technology

As discussed, privacy is a major consideration in many countries, as most have adopted GDPR-like standards. Since the European General Data Protection Regulation went into effect in 2018, there has been debate over the compatibility of developing technologies such as blockchains with new data protection standards. There are various points of contention about how to comply with obligations as stated in the various privacy laws discussed in the previous section; some of which are currently being debated by regulators or in court [11]. In drafting contemporary data privacy laws and frameworks, most legislators have failed to consider blockchain technology and its innovative features, which can help mitigate privacy concerns [15].

The effective applicability of various privacy laws and obligations has increasingly contributed to the success of businesses [10]. Blockchain is a decentralized technology, wherein several parties/individuals are directly involved in the processing of personal data. From the perspective of regulations and business standpoints, compliance with data protection laws is a critical component for blockchain technology to gain participants' trust and usage. The development of blockchain technology has made it possible for different parties to collaborate on a single blockchain. Such collaborations make it possible for data to be subcontracted by the data controller and processing data. However, regardless of the parties handling personal data, as discussed, privacy laws have placed a direct obligation on both data controllers and processors [11]. Blockchain is believed to be highly secure, but it poses significant regulatory barriers to data privacy as most privacy law mandates the right for personal data to be forgotten [10]. Exercising data protection rights by individuals whose personal data is processed in a blockchain environment is a concern, as it is unclear how these rights can really be enforced [11]. The contention between blockchain technology and data privacy regulations includes varying opinions on anonymity and pseudonymity and their correlation with data protection and privacy laws, the identification of data controllers and data processors in different blockchain technology implementations, the implications of distributed blockchain networks on territorial jurisdiction, the constraints that cross-border data transfers can have, the application of criteria to determine valid reasons for personal data processing in blockchain use cases and the reconciliation of data preservation and transaction immutability in blockchain applications with the rights of individuals [15].

Blockchain attributes

Consensus mechanisms utilize a protocol at the group level to attain dynamic agreement within the group are discussed below [9].

Proof of Authority (POA) was originally coined by Ethereum co-founder *Gavin Wood*. It is primarily designed for use in permissioned networks by selecting transaction validators based on their reputation and identity within the blockchain. Interested validators must provide the necessary documentation to reveal their identities and reputation, and all validators must be known to each other with no anonymity allowed. It has been argued that POA is more secure than proof of stake (POS) for two primary reasons. First, the incentive to validate transactions and blocks honestly is tied to the validators' identities, thus any misconduct will result in reputational damage. Second, the requirement for a validator to not approve two consecutive blocks prevents the concentration of trust [14].

> *Proof of Reputation (POR)* is another concept that is also used as a consensus algorithm. It is an extension of the POA algorithm implemented with various variations and parameters to optimize its performance. However, at its core, the algorithm is straightforward: a node's reputation is calculated based on predetermined formulas, and once it has proven its reputation and passed verification, it can be elected as an authoritative node. These authoritative nodes then function like those in a POA network where they exclusively validate and sign blocks [14].
>
> *Proof of Stake (POS)* is a consensus mechanism used in blockchain technology. It operates by maintaining a list of *validators* who have staked a deposit. In order to become a validator, a user must submit a particular transaction that binds a predetermined amount of their coins in a validator deposit. Validators can be identified by their Ethereum addresses, and the network chronicles all of the approved validators (those who have reserved coins for taking part in validation) [14].

A centralized blockchain grants the creator or owner complete authority over users' data. If a nation-state were to establish a centralized blockchain, it could be a concerning scenario for its citizens. It is probable that China will be the first country to adopt this approach as it has already implemented a closely supervised and Chinese-only internet, as well as its national digital currency. If Chinese citizens were obligated to use this blockchain, it could jeopardize the concept of self-sovereign identity that other blockchains potentially offer [2].

The blockchain is a secure distributed ledger that is immutable and verifiable. It works by grouping transactions into blocks that are then linked together in a chain. If a transaction is changed in one block, all subsequent blocks must be updated, making it difficult to alter any transaction on the blockchain. Although the blockchain provides security, concerns arise when personal data from IoT devices is gathered, stored and interpreted. These data are frequently stored in cloud computing environments, which creates more challenges for data protection and privacy regulations [20]. Despite its security features, records published on the blockchain cannot be deleted. Modern privacy legislation provides individuals with the *right to be forgotten*, but this right becomes challenging when dealing with permanent records on the blockchain. Therefore, it is necessary to find a way for individuals to exercise their right to be forgotten without compromising the security of the blockchain [21]. The collection and interpretation of personal data from IoT devices have raised significant privacy concerns, making IoT-based technology a critical priority for privacy and data protection regulators. Therefore, it is crucial to address the privacy implications of storing personal data on the blockchain and take appropriate measures to protect individuals' data.

Public blockchains are designed to be decentralized on purpose. This means that there isn't any one person or entity responsible for them. Such blockchains are made up of networks that can span across different locations (jurisdictions) and may involve many individuals who have the power to update the blockchain. This power is similar to that of a manager making decisions. However, since the responsibility for maintaining, managing and developing the blockchain is spread out among different unconnected people, regulators may find it challenging to take any action against the supporters of a public blockchain.

Blockchain implementation and privacy

Transparency is one of the main ways in which a blockchain ensures its security by permitting all involved parties with access to the blockchain to view the full record and history of transactions that have taken place. This enables each participant to verify that the information stored on the blockchain is accurate. However, if the information being stored on the blockchain contains personal information as defined under various privacy laws, this raises concerns about compliance with regulatory obligations. Compliance with certain obligations imposed on organizations that collect, use or disclose personal information can prove to be challenging, due to the core features of a blockchain such as immutability, transparency and accountability.

Public blockchains are intentionally designed to be decentralized, with no single entity being responsible for the network. This means that responsibility for responsibility of maintenance, administration, upkeep and progress is shared among a group community of unaffiliated individuals. While this approach provides benefits such as greater transparency, it can also pose enforcement challenges for regulators. In the context of Bitcoin, all transaction data is visible to users on the network. However, privacy can be maintained to some extent by breaking the flow of information in the transaction processing chain. Although the public can observe the amount being transferred, there is no information linking the transaction to any specific user. For optimal privacy, it is recommended to use a new key pair for each transaction to prevent transactions from being associated with any particular user. Nonetheless, multi-input transactions can reveal that the inputs were owned by the same person, thereby reducing privacy. As mentioned before, pseudonymous addresses (public keys or their hashes) offer partial unlinkability and provide some level of privacy. However, they are not entirely secure and can be compromised through various methods, limiting their ability to offer complete anonymity [22].

Privacy requirements of blockchain

Maintaining user privacy and security is a top priority in financial online transactions, particularly in blockchain applications. Users often prefer to limit the disclosure of their account information and transaction details within an online trading system. Achieving this minimal disclosure involves several measures, including preventing unauthorized access to transaction information, prohibiting system administrators or network participants from divulging user information without consent and ensuring secure storage and consistent access to user data even in the face of unexpected failures or cyberattacks.

Sharing user data between financial institutions can be expensive and difficult, leading to repeated user authentication. This can also increase the risk of identity disclosure by intermediaries. Additionally, in some cases, parties may be hesitant to reveal their true identity to one another, highlighting the importance of anonymity.

Privacy good practices and recommendations

In Canada, like many other countries, there are currently no official guidelines or interpretations regarding the processing of personal data on public or private blockchains. Under Canadian law, the wide range of personal information defined could raise the possibility of discouraging blockchain stakeholders from processing personal data on public blockchains. In the private blockchain context, individual rights over personal information can be effectively managed by assigning responsible entities with the power to control access to the blockchain. Stakeholders may also be required to comply with applicable privacy regulations in order to gain access to the private blockchain and its accompanying applications. Failure to comply with these regulations may result in removal from the network. Furthermore, a sufficiently centralized private blockchain may be vulnerable to being overwritten by participants in reaction to any privacy-infringing incidents. To proactively mitigate privacy law risks, stakeholders behind decentralized applications (DApps) in both public and private blockchain contexts can design appropriate privacy policies and implement best practices. These measures may include:

Combining On-Chain and Off-Chain Data can be done by ensuring that blockchain applications avoid storing identifiable information as a data payload on the blockchain, which means getting rid of personal information from the payment message. As a result, blockchain transactions should serve primarily as access control mechanisms or pointers guiding users toward more efficiently managed off-chain storage solutions. This approach will enhance clarity and professionalism in blockchain operations [21].

Utilizing Privacy Centric Technologies and cryptographic Methods through encryption techniques that prioritize user privacy, such as *zero-knowledge Succinct Non-interactive Argument of Knowledge (ZK-SNARKS), Ring Confidential Transactions* and *mixing techniques*, have become increasingly popular in privacy-centric chains. These techniques aim to mask the identities of senders and recipients and provide participants with cryptographic proof of transactional legitimacy without revealing sensitive information [21]. One particularly effective technique for preserving privacy is *zero-knowledge proofs*, which was initially proposed in the early 1980s. The concept involves formulating a formal proof that verifies the execution of a program with private input to produce publicly open output without disclosing any additional information. By using zero-knowledge proofs, a certifier can prove the accuracy of an assertion to a verifier without revealing any useful information. A more efficient variant of zero-knowledge proof of knowledge proof is zk-SNARK, which was introduced in 2012 by *Bitansky* and his coauthors and serves as the backbone of the Zcash protocol. *Zcash* uses zk-SNARKs to verify transactions while protecting users' privacy. To ensure a seamless and professional flow, privacy-centric chains should continue to prioritize the use of encryption techniques such as zero-knowledge proofs and zk-SNARKs. By doing so, these chains can provide participants with a secure and private transactional experience without sacrificing the efficiency and convenience of traditional blockchain technology.

Use of *Homomorphic Encryption (HE)* is a cryptographic technique that performs calculations on encrypted data and gets the same results as if performed on plaintext. It can be used to keep data private on public blockchains while still allowing authorized access for auditing and other purposes. Ethereum smart contracts use HE to keep data secure on blockchains.

Use of *Attribute-based Encryption (ABE)* is a cryptographic technique where the attributes of data are used to determine who can access it. The data is encrypted using a user's secret key and can only be decrypted by users whose attributes match those of the encrypted data. ABE is designed to be collusion-resistant so that even if a malicious user works with others, they cannot access data that they are not authorized to decrypt. Since its initial proposal in 2005, ABE has been extended to include schemes with multiple authorities and support for arbitrary predicates. Despite its power, ABE has not been widely deployed due to a lack of understanding and efficient implementation. However, a decentralized ABE scheme has been proposed for use on a blockchain. In this context, access to data can be represented by ownership of access tokens that are algorithmically distributed by a trusted authority. Multiple authorities can be used in a decentralized network, with witnesses potentially fulfilling this role. Although there are technologies, such as Steemit, Storj, IPFS and SAFE Network, that could enable an ABE implementation on a blockchain, it remains a challenge.

Trusted Execution Environment (TEE)-Based Smart Contracts involve an execution environment that is isolated from interference and information access from other software applications and operating systems, which is known as TEE. Intel Software Guard Extensions (SGX) is an example of TEE technology, while *Ekiden* is an SGX-based solution for smart contracts that maintain confidentiality. Ekiden separates computation from consensus by performing smart contract computation off-chain on compute nodes within TEEs. *Enigma* utilizes TEE to enable the implementation of a remote attestation protocol, which can be used to verify the correctness of compute nodes' execution on-chain. Moreover, consensus nodes are able to sustain the blockchain without having to rely on trusted hardware. With this technology, users are able to formulate privacy-preserving smart contracts that are powered by a decentralized credit scoring algorithm that takes into consideration various factors, including but not limited to, payment history, credit utilization, and the number and variety of accounts [14].

As discussed, Bitcoin transactions are publicly visible and can be traced back to the addresses used, potentially revealing a user's identity and compromising their privacy and security. To avoid this, users can utilize *mixing services*, which randomly exchange their coins with other users' coins to obfuscate their ownership [14]. However, mixing services have limitations such as a delay in reclaiming coins, potential coin tracing and theft by the mix operator, which may not be ideal for users seeking privacy protection. While users with something to hide may be willing to accept the risks associated with mixing, most users prefer to keep their spending habits private. Unfortunately, many users are unaware that their privacy has been compromised and many are unwilling to expend the continual effort necessary to protect their privacy. To address these concerns, users need a privacy protection mechanism that is instant, risk-free and automatic. Such a mechanism should prevent data revealing their spending habits and account balances from being publicly accessible to their peers, co-workers and merchants. Therefore, there is a need for more robust privacy protection mechanisms that do not suffer from the limitations of mixing services. These mechanisms should prioritize the user's privacy and security, and provide a hassle-free way to protect their anonymity on the blockchain [23].

Bitcoin Extensions or Altcoins refer to digital currencies that are modeled after Bitcoin. For example, *ZeroCoin* is a cryptographic extension to Bitcoin that enables anonymity by using zero-knowledge proofs to confirm encrypted transactions as valid [22]. ZeroCoin addresses some privacy concerns by dissociating transactions from their sources, but it still reveals payment destinations and amounts and its functionality is limited. Instead of relying on a central bank, Bitcoin its blockchain to record user transactions. The blockchain is publicly accessible as it is widely duplicated by untrusted peers. However, unlike a fully anonymous currency, ZeroCoin is a decentralized mix that allows users to periodically "wash" their bitcoins using the ZeroCoin protocol [23].

Anonymous Signatures such as digital signature technology that includes several variants, some of which provide anonymity for the signer, known as anonymous signature schemes. Two of the most widely utilized and characteristic anonymous signature schemes are group signatures and ring signatures. *Group signatures*, proposed in 1991, allow any member of a group to sign a message anonymously while only revealing the membership of the group during verification. A group manager is responsible for adding members, handling disputes and revealing the original signer's identity if necessary. In blockchain systems, an authority entity performs these functions by creating and revoking groups, dynamically adding new members and removing participants from the group. *Ring signatures*, on the other hand, achieve anonymity through a ring-like structure allowing any member of a group to sign a message anonymously. In the event of a dispute, ring signatures safeguard the real identity of the signer as there is no group manager in ring signatures. Without any additional setup, users can form a ring, making it appropriate for public blockchains. It is frequently applied in *CryptoNote*, where it conceals the link between the sender's addresses of transactions [14]. In summary, both group signature and ring signature are important anonymous signature schemes that provide a high degree of anonymity for the signer. Group signature requires a group manager and is suitable for private blockchains, while ring signature applies to public blockchains and does not require a group manager.

CONCLUSIONS AND RECOMMENDATIONS

This chapter provides an overview of the privacy concerns related to blockchain technology and emphasizes the importance of protecting consumer data privacy. The implementation of privacy laws and regulations has increased worldwide, and personal data protection acts have been established in various countries to regulate the collection, use, disclosure and storage of personal data. However, the decentralized nature of blockchain technology presents significant challenges to data privacy regulations. Therefore, maintaining user privacy on blockchain requires several measures, including preventing unauthorized access, protecting user data, and ensuring anonymity and unlinkability of transactions.

To address these challenges, organizations and blockchain stakeholders should:

- Stay updated with the latest privacy laws and regulations.
- Appoint privacy officers, and implement privacy protection measures. They should also provide employees with training on data protection, conduct regular risk assessments, establish data protection policies and procedures, and seek consent from data owners before collecting and using their personal information.
- Organizations should prioritize the use of encryption techniques such as zero-knowledge proofs like zk-SNARKs and consider using attribute-based encryption and trusted execution environment-based smart contracts.
- Organizations should also continuously review and update their privacy policies to meet evolving privacy law requirements.
- Governments should ensure effective enforcement mechanisms are in place to promote compliance with privacy regulations and continue to review and update these laws to address emerging privacy concerns, technological advancements and global best practices.
- Governments should also reconcile the competing interests of blockchain technology and data privacy regulations. This requires a continued dialogue between regulators and blockchain stakeholders to ensure that the interests of both parties are protected.

In conclusion, privacy and data protection are crucial considerations in the modern age, particularly with the rise of technology and the digital economy.

REFERENCES

[1] Pala, K. (2021, September 8). *Privacy, security and policies: A review of problems and solutions with blockchain-based internet of things applications in manufacturing industry*. Procedia Computer Science. Retrieved November 22, 2022, from https://www.sciencedirect.com/science/article/pii/S1877050921014174

[2] Piwik PRO. (2022, October 12). 11 new privacy laws around the world and how they'll affect your analytics. *Piwik PRO*. Retrieved November 17, 2022, from https://piwik.pro/privacy-laws-around-globe/

[3] GeeksforGeeks. (2022, May 30). Blockchain and Data Privacy. *GeeksforGeeks*. Retrieved November 18, 2022, from https://www.geeksforgeeks.org/blockchain-and-data-privacy/

[4] Bennett, K. (2021). *Blockchain Overview*. Blockchain Training Alliance. Retrieved November 21, 2022, from https://courses.btacertified.com/courses/take/blockchain-basics/pdfs/36187873-slides

[5] IMI https://www.identitymanagementinstitute.org/app/uploads/2021/03/logo-.jpg. (2021, May 3). *Blockchain data privacy concerns*. Identity Management Institute®. Retrieved November 22, 2022, from https://identitymanagementinstitute.org/blockchain-data-privacy-concerns/

[6] Walters, N., & Coutu, S. (2022, June 9). *The privacy paradox in Blockchain: Best practices for data management in Crypto*. Dentons. Retrieved November 22, 2022, from https://www.dentons.com/en/insights/articles/2022/june/9/the-privacy-paradox-in-blockchain-best-practices-for-data-management-in-crypto

[7] SNIA. (n.d.). What is data privacy? *SNIA*. Retrieved October 12, 2022, from https://www.snia.org/education/what-is-data-privacy#:~:text=Data%20privacy%2C%20sometimes%20also%20referred%20to%20as%20information,protecting%20the%20confidentiality%20and%20immutability%20of%20the%20data.

[8] Tapscott, D., & Tapscott, A. (2018). *Blockchain Revolution: How the technology behind Bitcoin and other cryptocurrencies is changing the world*. Penguin, an imprint of Penguin Canada.

[9] Conti, M., Kumar, S. E., Lal, C., & Ruj, S. (2018, May 31). *A survey on security and privacy issues of bitcoin*. IEEE Xplore. Retrieved November 22, 2022, from https://ieeexplore.ieee.org/document/8369416

[10] GDPR. (2022, May 26). What is GDPR, the EU's new Data Protection Law? *GDPR.EU*. Retrieved November 12, 2022, from https://gdpr.eu/what-is-gdpr/

[11] GDPR EU. (2021, March 8). GDPR personal data – what information does this cover? *GDPR EU*. Retrieved November 19, 2022, from https://www.gdpreu.org/the-regulation/key-concepts/personal-data/

[12] DelaCruz, R. (2019). Privacy laws in the blockchain environment. *AETiC Special Issue on Next Generation Blockchain Architecture, Infrastructure and Applications*, 3(5), 34–44. https://doi.org/10.33166/aetic.2019.05.005

[13] Torks, A. (2021, October 4). What's the difference: Anonymity vs. pseudonymity. *Utopia. Fans*. Retrieved November 14, 2022, from https://utopia.fans/privacy/whats-the-difference-anonymity-vs-pseudonymity/

[14] Batmunkh, D. (2018, November 28). *Private blockchain consensus mechanisms*. Medium. Retrieved November 21, 2022, from https://medium.com/@arigatodl/private-blockchain-consensus-mechanisms-8e6fc48c8fb

[15] Kamalpour, A., Shivarattan, G., & Mokhtassi, I. (2020, December 9). Privacy laws and Blockchain. *Ashurst*. Retrieved November 17, 2022, from https://www.ashurst.com/en/news-and-insights/insights/cityam-talking-legal---privacy-laws-and-blockchain/

[16] Certik. (2022, April 21). What is pseudonymity and anonymity? – blog – certik security leaderboard. What is Pseudonymity and Anonymity? - Blog - CertiK Security Leaderboard. Retrieved November 14, 2022, from https://www.certik.com/resources/blog/what-is-pseudonymity-and-anonymity

[17] Zhang, R. U. I., Xue, R., & Liu, L. (2020, May 1). *Security and privacy on Blockchain*. ACM Computing Surveys. Retrieved November 21, 2022, from https://dl.acm.org/doi/10.1145/3316481

[18] Dulberg, R. (2021, September 13). Why the world's biggest brands care about privacy - medium. *UX design*. Retrieved November 19, 2022, from https://uxdesign.cc/who-cares-about-privacy-ed6d832156dd

[19] Ben-Sasson, E., Chiesa, A., Garman, C., Green, M., Tromer, E., & Miers, I. (2014, May 18). *Zerocash: Decentralized anonymous payments from Bitcoin (extended version)*. Zerocash: Decentralized Anonymous Payments from Bitcoin. Retrieved November 21, 2022, from http://zerocash-project.org/media/pdf/zerocash-extended-20140518.pdf

[20] PDPC Singapore. (n.d.). *PDPC: PDPA Overview*. Personal Data Protection Commission. Retrieved November 17, 2022, from https://www.pdpc.gov.sg/Overview-of-PDPA/The-Legislation/Personal-Data-Protection-Act

[21] Centers for Disease Control and Prevention. (2022, June 27). *Health Insurance Portability and Accountability Act of 1996 (HIPAA)*. Centers for Disease Control and Prevention. Retrieved November 17, 2022, from https://www.cdc.gov/phlp/publications/topic/hipaa.html

[22] Office of the Privacy Commissioner of Canada. (2019, May). *The Personal Information Protection and Electronic Documents Act (PIPEDA)*. Office of the Privacy Commissioner of Canada. Retrieved November 17, 2022, from https://www.priv.gc.ca/en/privacy-topics/privacy-laws-in-canada/the-personal-information-protection-and-electronic-documents-act-pipeda

[23] PIPL. (2022, May 10). Personal information protection law of the People's Republic of China. *PIPL*. Retrieved November 17, 2022, from https://personalinformationprotectionlaw.com/

Chapter 5

Blockchain security considerations

Lilian Behzadi and James Joseph

BACKGROUND

Blockchain technology started in 1991 and was initially only used for financial (cryptocurrency) transactions. In 2014, other options were explored, and the most recent advancements and potentials were seen in other financial and inter-organizational areas. During the past few years, blockchain usage has multiplied across all industries for various deployments and use cases. Businesses currently use blockchain technology to manage distributed databases, digital transactions, cybersecurity and healthcare to create blockchain-based solutions for clients. As mentioned in previous chapters, the main advantage of using blockchain technology is that it guarantees transaction security due to its cryptographic, decentralized and consensus-based features. According to a recent report, the worldwide blockchain market is expected to rise to $20 billion by 2024. Currently, 69% of banks are looking into various blockchain technology options to improve the security, consistency and simplicity of services. While the adoption of blockchain offers numerous benefits for businesses globally, the technology has also drawn a lot of attackers who want to hack systems and cause harm to companies. In the domain of cyberattacks and hacking attempts, blockchain security has recently grown to be a critical component of procedures that helps organizations and businesses to safeguard operations [1].

Data structures created by blockchain technology include built-in security features. The structures are founded on cryptographic, decentralized and consensus mechanisms that guarantee the integrity of transactions. The data is organized into blocks in blockchains or *Distributed Ledger Technologies (DLT)*. Each block contains a transaction or a collection of transactions. In a cryptographic chain, each new block is connected to all the blocks that came before it, making tampering with the blocks nearly impossible. A consensus mechanism verifies and accepts each transaction contained within the blocks, ensuring that each transaction is accurate and true [2].

Blockchain technology relies on CIA pillar: Confidentiality, Integrity and Availability. Confidentiality provides a wide variety of capabilities to ensure a user's anonymity. Public and private keys are the only link between users and their data. Some networks on the blockchain use non-interactive zero-knowledge proofs to maximize a user's confidentiality. Integrity is where each block on the chain is linked to other blocks using cryptographic hash functions. Once a transaction is recorded on the blockchain, it can't be modified or deleted. Any changes or modifications must be processed as new transactions. Availability is where many nodes on the blockchain ensure resilience even when some nodes are unavailable. The correct blockchain remains accessible to other users even with a compromised node. This is because each node in the network has a copy of the distributed ledger [3].

The *Open Web Application Security Project (OWASP)* is a non-profit organization dedicated to enhancing software security. The OWASP project aims to improve security by providing information to developers, including detailed lists of common vulnerabilities and resources, such as testing tools and vulnerable applications to help educate developers about the various security risks that could exist in written code. blockchain technology is quickly gaining popularity, but currently there aren't many related tools or security standards like the *OWASP* list. A similar top ten list produced by the *Decentralized Application Security Project (DASP)* is focused on smart contract vulnerabilities and does not consider the entire blockchain ecosystem [4].

In 2017, the OWASP top ten list of web application vulnerabilities was published with ten common vulnerabilities affecting application security. The common vulnerabilities included *Injection, Broken Authentication, Sensitive Data Exposure, XML External Entities (XXE), Broken Access Control, Security Misconfiguration, Cross-Site Scripting (XSS), Insecure Deserialization, Using Components with Known Vulnerabilities, and Insufficient Logging and Monitoring.* The list was used for web applications, but many vulnerabilities are also relevant to blockchain technology [4]. The OWASP top ten List was updated in 2021 with changes to the common vulnerabilities that still affect application security. The list of vulnerabilities now includes Broken Access Control, Cryptographic Failures, Injection, Insecure Design, Security Misconfiguration, Vulnerable and Outdated Components, Identification and Authentication Failures, Software and Data Integrity Failures, Security Logging and Monitoring Failures and Server-Side Request Forgery (SSRF) [5], as shown in Table 5.1.

DISCUSSION

Four attributes make a blockchain trustworthy. These include consensus, non-modifiability, cryptography and decentralization. Consensus is the ability of nodes gathered within the network to agree on one true network state and transaction validity. This depends on the type of algorithm used, like *Proof of Work (PoW)* and *Proof of Stake (PoS)*. Non-modifiability is the ability of a network to prevent changes to transactions that have already been approved (Finality) [16]. The concept of *Finality* ensures that transactions can't be altered, terminated or reversed after they are approved. Finality measures the time it takes for an individual to receive a reasonable guarantee that an attacker will not tamper with the transactions. The *latency level* (time delay) ultimately affects the blockchain's finality rate. It is crucial for organizations to have a low latency when accepting crypto as a means of payment because waiting for a long time on a blockchain network for a transaction to clear can often have adverse effects on organizations [17]. Any attempts to do any modifications or conduct unauthorized transactions on the blockchain will be detected and automatically rejected. Any changes to blockchain transactions are impossible without leaving a trace. All the transactions within the blockchain are secured by using cryptographic functions. Each block has a unique private key that is verified with a public key. If the data within a transaction changes, the key becomes invalid and that block is soon removed from the chain. Decentralization is where no single entity is needed. This includes a single computer or an infrastructure that would be vulnerable to attacks or other issues [16].

Attacks at the four vector levels

An *attack vector* is an entry point a hacker uses to gain unauthorized access to launch a cyberattack. Attackers can acquire sensitive data, personal information and other valuable information available after a data breach by using attack vectors to take advantage of system flaws. It is important for companies to implement mitigation techniques to minimize potential attack vectors and prevent data breaches. Infrastructure, computer systems, networks, operating systems and IoT devices can all be exposed, altered, disabled, destroyed, stolen or

Table 5.1 2021 OWASP top ten vulnerabilities

Vulnerability	Description
A01 Broken Access Control	By enforcing policy and regulations, access control ensures that users stay within the bounds of the specified permissions. Failures frequently result due to unauthorized access, disclosure of information, modification or deletion of data, or execution of business operations outside the user's scope [6].
A02 Cryptographic Failures	Identifying data protection requirements both in transit and at rest is critical. Passwords, credit cards, health records and personal information need extra protection, especially if the data is covered by privacy laws like *General Data Protection Regulation (GDPR)* or rules like *PCI Data Security Standard (PCI DSS)* [7].
A03 Injection	An application is exposed to an attack if an application does not check, filter or sanitize user-supplied data. Object-relational mapping (ORM) search parameters are employed in conjunction with hostile data to retrieve more sensitive records. Dynamic queries and commands contain harmful data structures [8].
A04 Insecure Design	Missing or ineffective control design refers to a range of flaws known as insecure design. A correctly implemented secure design may still have flaws that can be exploited. However, a perfect implementation can't fix an unsafe design as developers have not created and implemented the necessary safeguards to protect against attacks [9].
A05 Security Misconfiguration	An application might be vulnerable if appropriate security hardening is missing, permissions are not configured properly, unnecessary features are installed and the latest security features are disabled or not configured securely [10].
A06 Vulnerable and Outdated Components	Systems are vulnerable if the components' versions on the client and server sides are unknown, the software is not updated and unsupported, and platforms and frameworks are not fixed and upgraded regularly [11].
A07 Identification and Authentication Failures	It is essential to confirm a user's identity, authenticate user information, and monitor and manage sessions to safeguard against threats linked to authentication. Authentication weaknesses result from brute force, automated attacks like credential stuffing, weak passwords, and ineffective multi-factor authentication [12].
A08 Software and Data Integrity Failures	Code and architecture that do not protect against integrity violations are related to software and data integrity failures. For example, the auto-update functionality poses a threat as updates are downloaded without sufficient verification and applied to trusted applications [13].
A09 Security Logging and Monitoring Failures	The category is used to identify, prioritize and address current breaches. Breach detection is impossible without logging and monitoring activities. Inadequate logging, detection and monitoring results in failures [14].
A10 Server-Side Request Forgery	When a web application retrieves a remote resource without checking the user-supplied URL, SSRF issues can occur. Even when a firewall, VPN or network access control list is implemented to safeguard an application, SSRF enables an attacker to force the application to submit a forged request to an unexpected location [15].

subjected to other types of unauthorized access in various ways. Attack vectors can be split into active or passive attacks. An active attack vector aims to alter a system or modify how the system functions. Passive attack vector exploits are attempts to access or use data from the system without depleting system resources [18].

Blockchain attacks occur in a similar way at four different vectors: system level, network level, user level and smart contracts. System-level attacks are where hackers look for vulnerabilities in the operating system to gain access to the target system. The hackers exploit the vulnerabilities, which allow access to the system's open ports and services. The open ports and services are common vulnerabilities in the operating system and are installed by default [19].

Attackers at the network level get unauthorized access to an organization's network. The aim is to steal confidential information and data or perform malicious activities. Passive attacks and active attacks can take place at the network level. To reiterate, a passive attack occurs when an attacker gains access to the network and monitors or steals sensitive data without making any changes to the data and leaving it intact. An active attack occurs when an attacker gains unauthorized access to the network and modifies the data by either encrypting, deleting or doing something harmful [20].

Attacks at the user level happen when attackers misuse permissions to gain access to data for transactions. The users are broken up into four unique categories. The first category is blockchain users, who are participants with permission to join the network and conduct transactions with other users. The second category is regulators, who are users with special permissions to oversee transactions happening within the network. The third category is blockchain network operators, who are users with special permissions and have the authority to define, create, manage and monitor the network. Finally, the fourth category is certificate authorities, who are individuals that issue and manage different types of certificates. The certificates are required to run a permissioned blockchain [21].

Smart contracts are programs or lines of code on a blockchain that run when predefined conditions are met. Smart contracts are used to automate the execution of agreements between different parties so that the parties involved are aware of the outcome without needing a mediator. Attacks on smart contracts are used to falsify smart contract rules. The code and agreements made by smart contracts exist across a decentralized blockchain network. The code controls the execution of each transaction and the transactions are trackable and irreversible. There are a variety of different blockchain-related attacks that occur at the user level, network level, system level and smart contracts [22].

User-level attacks

Some user-level attacks include *Digital Signature Vulnerability*, *Hash Function Vulnerability*, *Finney Attacks*, *Vector76 Attacks*, *51% Attacks* and *Phishing Attacks*. The Digital Signature Vulnerability is based on asymmetric cryptography that uses public and private keys, making it hard for attackers to duplicate. The use of algorithms generates both public and private keys. The private key is only known to the signing party and stored on the party's device or the cloud for centralization. The public key is shared among recipients of the digital signature process. The digital signature creation is done when the algorithm yields the private key and a corresponding public key. Given a message and a private key, a signing algorithm then generates a signature that is encrypted by the private key. The digital signature can't be forged without access to the private key. An algorithm used to verify signatures or documents either accepts or rejects the authenticity claim of the message based on the message itself, the public key and the signature. The main motive of the attack is to obtain the customer's private key that is used to sign messages or documents. Attackers take advantage of the missing steps in the validation process and not on signature generation [23]. An incident where the *Digital Signature Vulnerability* was exploited was when an attack campaign manipulated the *Microsoft* signature verification that targeted 2,100 customers. The attackers deployed a banking malware called *Zloader* onto the system. Zloader was used to steal user credentials and private information. The attack campaign took place in November 2021, and results showed that a total of 2,170 IP addresses worldwide had downloaded the malicious dynamic-link library (DLL) file [24].

The *Hash Function Vulnerability* is a type of attack where an attempt is made to find two strings of a hash function that produce the same hash result. A collision occurs when two pieces of data (e.g., binary, document or a website certificate) hash to the same digest. If the hash algorithm, that is, SHA-1, has flaws, an attacker can cause a collision to occur. The attacker can use the collision to deceive systems into accepting a malicious file. An example of the occurrence of a collision is two insurance contracts with significantly different terms and conditions. The likelihood of a *Collision Attack* occurring is extremely low, particularly for functions with huge output sizes like lengthy and popular formats or protocols. Attacking hash functions is becoming easier as computing power becomes more accessible and affordable. In 2017, *Google* revealed that a group of researchers from the *CWI Institute* in Amsterdam and *Google* had partnered and successfully performed an attack on the *SHA-1* hash algorithm by creating two files that hashed to the same value [25].

Hal Finney was the first to receive a Bitcoin transaction and the first to announce the creation of *Bitcoin*. Finney was also the first to raise the possibility of a Bitcoin double-spending attack. The attack was given the name *Finney Attack* or *Finney Hack* in his honor. A Finney Attack is a type of double-spending attack that happens when a person accepts an unconfirmed transaction on the network. It should be noted that this type of double-spending attack is difficult to execute. It is challenging to perform the attack because the attacker is a miner capable of extracting the block where his transaction will be approved. Moreover, a merchant must agree to accept a transaction with no network confirmations. The two requirements are tough to combine to make the attack successful.

To explain further, in a Finney Attack, the double-spending scenario unfolds as follows:

The attacker (who is also a miner) initiates a transaction from Address A to Address B within a block that they mine. Simultaneously, the attacker sends another transaction from Address A to Address C, belonging to a different user. The second user at Address C, unaware of the initial transaction, accepts the payment without waiting for network confirmation. Subsequently, the attacker releases the mined block containing the initial transaction from Address A to Address B. This action invalidates the transaction to Address C, allowing the attacker to double spend on behalf of the second user. As such, the key to the attack lies in the miner's ability to control the release of the block manipulating the sequence of transactions and taking advantage of the lack of confirmation from the network for the second transaction [26].

The *Vector76* Attack is a type of double-spending attack that takes advantage of minor bugs in the consensus system of *Bitcoin*. The target of the Vector76 Attack is exchanges or exchange houses, where attackers can buy and sell cryptocurrency and tokens without being quickly discovered [27]. The attack occurs when a miner controls a network with two full nodes and connects the first node (node A) directly to the exchange service. The second node (node B) connects to the other nodes located on the network. The attacker tracks the time at which the nodes transmit transactions and how they can propagate the transactions to other nodes. After establishing a connection, the miner generates a valid block. A pair of transactions is created with different values: a high-value transaction and a low-value transaction. The attacker sends the low-value transaction, which rejects the high-value transaction, and as a result, the attacker is credited the amount of the high-value transaction [28].

The *51% Attack* is a top attack that happens when a group of miners own more than 50% of the blockchain network's mining hash rate. The parties involved get control to alter the blockchain. Owning 50% of the nodes gives the parties involved the power to modify the blockchain and make changes to transactions. Attackers can prevent new transactions from getting confirmed, not allowing users to get paid, and they can reverse transactions that are already verified. This attack can result in users doubly spending [29]. An example of where a *51% Attack* occurred was with *Bitcoin Gold (BTG)*. The attack resulted in $72,000 worth of BTG tokens being double spent. The attack was split over two days. On the first day, the attackers removed 14 blocks from the *BTG* network, replacing them with 13 blocks and double spending 1,900 BTG worth around $19,000 at the time. On the second day, the attackers removed 15 blocks from the network and issued another 16 blocks, resulting in a double spend of 5,267 BTG, approximately over $53,000 [30].

A *Phishing Attack* is a type of cryptocurrency scam that involves deceiving users into disclosing private keys or personal information. The attacker frequently poses as a legitimate entity or individual to win the trust of the user. The attacker uses the user's details to steal the cryptocurrency funds after the victim has been scammed. An attacker begins a Phishing Attack by sending out a mass email or message to potential victims. The email or the message appears to be coming from a reliable source like a wallet or cryptocurrency exchange. A link to a fake website that looks identical to the actual website is always included in the message. The attacker misuses the user's login credentials when the link is clicked and gets authorized access to the user's account, files and other sensitive information [31]. An incident where a *Phishing Attack* took place was on August 4, 2022. *Twilio*, a digital communications platform was hacked after a phishing campaign tricked company employees into disclosing login credentials. The attack was a *Social Engineering Attack*, which is a type of *Phishing Attack*. The company reported the breach on the *Twilio* blog three days after the attack. A limited number of customer accounts were affected by the attack. Twilio enables web services to send SMS messages and make voice calls over telephone networks. These services are used by companies like *Uber*, *Twitter*, and *Airbnb*. The hack involved an attacker sending SMS messages to Twilio employees to reset passwords or alerting the employees to a change in their schedule. Each message contained a link with keywords like *Twilio*, *SSO (single sign-on)*, and *Okta*, the name of the user authentication service that is utilized by many organizations. Employees who clicked on the link were directed to a fake *Twilio* sign-in page, where hackers were able to capture the data entered into the fields. *Twilio* worked with phone carriers to shut down the SMS scheme and requested that web hosting companies remove the fake sign-in pages [32].

Network-level attacks

Network-level attacks include Side Chain Attacks, Sybil Attacks, Eclipse Attacks, Transaction Malleability Attacks, Timejacking Attacks and Distributed Denial of Service (DDoS) Attacks. The *Side Chain Attack* is an attempt to extract secrets from a chip or a system. The extraction is achieved by measuring or analyzing physical parameters like execution time, supply current and electromagnetic emission. The *Side Chain Attack* doesn't directly target a system or the code but tries to gather information and manipulate the program execution of the system by measuring or exploiting the indirect effects of the system or its hardware [33]. An incident where the Side Chain Attack took place on the *Binance Cryptocurrency Exchange*. An attacker stole $570 million of crypto tokens from Binance's BNB Chain. The attack hit the *Binance Bridge*, the cross-chain bridge that allows for the transfer of tokens between two related blockchains operated by the *Binance Cryptocurrency Exchange*. The attacker forged transactions that allowed the withdrawal of 2 million *BNB* tokens from the bridge worth $570 million [34].

A *Sybil Attack* is arranged by assigning several identifiers to the same node. There are no trusted nodes on the networks as a malicious node claims fake identities or generates new identities. The attacker controls multiple nodes, and the user is surrounded by fake nodes that close all their transactions [35]. Attackers have the potential to carry out unauthorized actions in the system if a *Sybil Attack* is successful. The attack allows a single entity like a computer to generate and manage many identities, such as user accounts and IP address-based accounts. All of these false identities deceive individuals and computer systems into thinking that the identities are real. The name *Sybil Attack* is taken from a 1973 novel about a lady named Sybil, who had dissociative identity disorder. The term was first used in relation to attacks by *Brian Zill* and *John R. Douceur*, who worked for the research department at *Microsoft* [36]. An incident where the *Sybil Attack* took place was in 2015 when a Swiss company, *Chainalysis*, had created over 250 false Bitcoin nodes to gather information on the whereabouts of transactions. The attack occurred when an individual created multiple fake identities to gain influence in a peer-to-peer network. The company created fake nodes to gather location data for a blog post about *Bitcoin* transfers between countries [37].

An *Eclipse Attack* is designed to separate a single node from the rest of the network. The separation requires that the attacker control the node's connection to all the other nodes in the network. The control of the node's connection is accomplished in a variety of ways by malicious neighbor nodes, malware and *Man-in-the-Middle Attacks*. Malicious neighbor nodes work in a way that every node on the network is directly connected to other nodes and receives transactions and blocks via those nodes. If an attacker controls all the neighbor nodes, then the attacker has control over the node's view of the blockchain. Blockchain software relies on computer hardware to perform tasks. The software and network connection communications can be filtered by malware that can intercept the communication. An attacker can take over a node's network connection by performing a *Man-in-the-Middle (MitM) Attack* via a Wi-Fi connection or Internet Service Provider (ISP). The Man-in-the-Middle Attack allows the attacker to filter traffic meant for other nodes. During an Eclipse Attack, the attacker filters the traffic going to the node from other nodes, enabling it to drop transactions or blocks. The attacker can also send conflicting versions of transactions or blocks to the target node and the rest of the network. The attacker achieves a *Denial of Service (DoS)*, divergent blockchain and a double spend [38].

The *Transaction Malleability Attack* is an attack that changes the unique transaction ID before it is approved or validated by the network. For attackers to alter transactional information, access is required to the blockchain network. The attackers gain unauthorized access to the network and change transactional details. If a transaction is modified, the modification indicates that an attacker could access the network successfully. Blockchain security becomes invalid if an attacker alters the information before generating a hash. An attacker can efficiently disregard any valid user transactions on the network. In some circumstances, blockchain security may disregard the transaction and the cryptocurrency wallet may continue to believe that no Bitcoin was sent in the first place. According to a number of analysts, the Transaction Malleability Attack created the conditions for double spending and *51% Attacks*. The attack makes a single coin expenditure and before the first transaction is confirmed, the attack manages to create other transactions using the same Bitcoin or other cryptocurrencies [39]. A major incident where the Transaction Malleability Attack occurred was in 2014 at a Japanese Bitcoin exchange called *Mt Gox*. *Mt Gox* was one of the biggest Bitcoin exchanges in the world. The company was handling over 70% of all Bitcoin transactions globally. By the end of that year, the company went bankrupt due to the massive hack it had suffered. Around 740,000 Bitcoins, or 6% of the total amount ever created, were lost by Mt Gox. The Bitcoins were worth over $3 billion in October 2017 and the equivalent of £460 million at the time. The company's bank accounts were short an extra $27 million. The company recovered 200,000 Bitcoins, but the remaining 650,000 Bitcoins were never found [40].

The *Timejacking Attack* exploits a vulnerability in the Bitcoin timestamp handling. An attacker compels a node to accept an alternate blockchain by changing the node's network time counter during a Timejacking Attack. The malicious attacker accomplishes the time change by adding multiple fake peers to the network with inaccurate timestamps. A poisoned block is created by the attacker manipulating one of the timers on the victim node. The rest of the network successfully accepts the block and the victim rejects the block creating a fork. The victim considers and accepts the alternative chain with token reuse (or double spending) valid that does not contain the "poisoned" block while the entire network uses the block-chain created for the poisoned block. The Timejacking Attack is carried out in two stages: fork and isolate, then double-spend. During the first stage (fork and isolate), the attacker needs to generate a new block, the same poisoned block that the user ignores. The user rejects the other blocks that are added after the poisoned block. The attacker manipulates the time value of the user's network by connecting to enough neighbors and broadcasting a lagging system time. The user's network time value decreases and the range of valid values of possible timestamps decreases. Accordingly, the poisoned block is a block whose timestamp must fall into the time window of all network nodes, except the users. The second stage (double spending) is where the attacker generates an alternative chain in which a transaction is added that transfers tokens to the user's wallet. Blocks of the alternative chain are sent to the user. The user accepts the chain since the value of the timestamps is chosen by the attacker and the remaining nodes are ignored. The result is the user believes that the tokens are received and sends money while the main network has record that the tokens did not leave the attacker's wallet and money comes to the attacker [41].

A *DDoS Attack* occurs when several systems coordinate a synchronized *DoS Attack* on a single target. The attack is a malicious attempt to disrupt the traffic of a targeted server, service or network by flooding the target with an excessive amount of traffic. DDoS attacks are carried out using a network of internet-connected devices. The networks are made of computers and other devices such as Internet of Things (IoT) devices that have been infected with malware, enabling an attacker to remotely manage the devices [42]. The main distinction between a DDoS Attack and a DoS Attack is the target is attacked simultaneously from multiple locations rather than a single location. A DoS attack aims to shut down a computer system or network so that the intended users are unable to access files and data. DoS Attacks achieve a system shutdown by flooding the target system or network with an excessive amount of traffic or information that causes a crash. The DoS Attack denies the service or resource that legitimate users such as employees, members or account holders expected. There are two methods used to execute DoS Attacks. The methods are flooding services or crashing services. Flood attacks occur when the system receives too much traffic for the server to buffer, causing the system to slow down and stop. A type of DoS attack is the DDoS Attack [43]. In 2017, *Bitfinex*, one of the largest cryptocurrency exchanges in the world, suffered a DDoS Attack. The attack was a second attempt in less than a week. As a result, Bitcoin prices fell 2% hitting $7,373.47. The attack only affected trading operations, and user accounts and associated funds/balances were not at risk during the attack. *Bitfinex* was founded in Hong Kong in 2012 as a peer-to-peer Bitcoin exchange. The company later added support for other cryptocurrencies [44].

System-level attacks

Attacks at the system level include Selfish Mining Attacks, Bribery Attacks, Crypto-jacking Attacks, Fork After Withholding Attacks and Malware Attacks. The *Selfish Mining Attack* (also known as the *Block Withholding Attack*) occurs when malicious miners withhold newly mined blocks. The attackers (miners) increase the share of the reward by not broadcasting mined blocks to the network. Instead, the miners keep the blocks a secret and start creating their own branches on the network. The miners then release several blocks at once, making other miners lose the originally published blocks. The malicious miners gain unfair revenue through the release of the blocks. The miners focus on a three-block strategy to make the attack successful. The mining continues on the private branch until the private branch is three blocks longer than the main branch. When a user (miner) finds the next block, the attacker publishes the private branch immediately. The attacker can then successfully claim the rewards due to the PoW protocol. The user's computing power is wasted, and no revenue is gained or return on investment [45]. In 2018, a Japanese cryptocurrency called *Monacoin* suffered a Selfish Mining Attack. The attack caused approximately $90,000 in damages. The attack was noted as a Selfish Mining Attack where one miner successfully mined a block on the blockchain and did not broadcast the new block to the other miners. The secret miner can successfully create a branch that is longer than the regular chain if a second block is found. According to most blockchain protocols, the chain with the most blocks is considered the correct chain, because it is the chain that has most of the PoW. When the longest chain becomes public, all the other blocks discovered by other miners become invalidated. In the case of *Monacoin*, the attacker sent the cryptocurrency to other exchanges outside the country. An example provided was *Livecoin*. The aim of the attack was to swap the cryptocurrency with other currencies before the hidden chains were publicly revealed. The miner had the computational power necessary to carry out the attack and took over 57% of the hash rate at one time. The attacker (miner) had been trying to take advantage of a flaw in how *Monacoin* changed its difficulty [46].

A *Bribery Attack* takes place when an attacker rents an amount of space at an approximate cost. Let's assume the cost of B per block mined, where B is the mining reward for one block. The attacker publishes a transaction T in block B and waits until other blocks have been published so that an irreversible action is taken because of T and introduces a new block B with the conflicting transaction T. The attacker then rents sufficient capacity to extend the branch containing B until it becomes the longest branch. The end result is that the attacker has double spent the funds in transaction T and can earn a profit equal to the value of T [47].

The *Crypto-jacking Attack* is a top attack that happens when an attacker abuses a user device's computing power with unauthorized access and uses it to mine cryptocurrency. The attack is delivered in the form of malicious software that infects devices used for mining. The target can be any device: computers, smartphones or cloud servers. When a device is infected with malware, it takes control of the device's computing power and controls a part of it to mine specific cryptocurrencies. The mined coins are finally sent to the attacker's wallet. The motive of this attack is to make a profit, but it is specifically designed to stay hidden from users [48]. An example of where the Crypto-jacking Attack took place was in 2018. *Tesla's* cloud system was hijacked by attackers who used it to mine cryptocurrency. The attackers could infiltrate the Kubernetes administration console because it was not password protected. *Kubernetes* is a *Google*-designed system that is used to optimize cloud applications. Access credentials to the company's *Amazon Web Services (AWS)* account were exposed, and the attackers deployed *Stratum*, a crypto mining software to mine cryptocurrency using the cloud's computing power [49].

The *Fork after Withholding Attack* is very rewarding to miners. The attack combines elements of the Selfish Mining Attack and Block Withholding Attack. The attacker, either an individual or belonging to a pool, splits the computing power between innocent mining and infiltration mining where infiltration mining is mining as part of a target pool. When the attacker discovers a *Full Proof of Work (FPOW)* as part of the infiltration mining in a pool, the attacker keeps the block private and does not publish it [50]. Depending on the situation, a malicious miner hides a winning block and either discards or releases the block to create a fork [3]. At the moment, there are not a lot of evaluation frameworks for *Fork After Withholding Attacks*, and there is a lack of revenue evaluation when the mining pool encounters a *Fork After Withholding Attack*. The assessment of *Fork After Withholding Attacks* in the evaluation mining pool and target mining pool is considered single, that is, one mining pool evaluates one target mining pool [51]. An incident where the *Fork After Withholding Attack* took place occurred when attackers drained almost $200 million in cryptocurrency from a tool called *Nomad*. The tool allows users to swap tokens from one blockchain to another. The attack highlighted weaknesses in the decentralized finance space. Blockchain security specialists referred to the hack as a "free for all." Any user or attacker who understood how the hack worked could have taken advantage of the weakness and used Nomad as an automatic cash dispenser, dispensing tokens at the touch of a button. The attack was successful due to an upgrade to Nomad's code. One part of the code was marked as valid, and initiating a transaction allowed attackers to withdraw more assets than were deposited onto the platform. Without programming knowledge, any user could copy the attacker's call data and substitute the attacker's address to exploit the protocol [52].

A *Malware Attack* is a common and top attack that occurs when unauthorized actions are conducted using malware (malicious software). The malicious software consists of different types of attacks, for example, *Ransomware Attacks*, *Spyware Attacks*, and *Command & Control*. The aim of a Malware attack is to exfiltrate information, disrupt operations and eventually demand payment. The main types of attack vectors used are worms, trojan horses and viruses [53]. A newly generated malware, the *Clop* ransomware, emerged in 2019 and has become an ongoing threat to organizations. The Clop ransomware encrypts files and threatens to leak information from those files if a ransom is not paid. Clop ransomware is a variant of *CryptoMix* Ransomware that encrypts data and renames every file by adding the *.clop* extension. A dangerous aspect of the ransomware is its ability to disable *Windows Defender* and remove *Microsoft Security Essentials*. The dangerous aspect helps the attacker gain unauthorized access to a victim's system [54].

Smart contract attacks

Attacks on smart contracts include Flash Loan Attacks, Parity Wallet Hack, and Distributed Autonomous Organization (DAO) Attacks. A *Flash Loan Attack* is a top attack that occurs when a platform's smart contract security is exploited. Flash Loan Attacks *are DeFi* platform attacks where a hacker borrows a significant amount of cryptocurrency through a flash loan. The hacker then uses the borrowed money to control the prices of certain assets on the platform [55]. The hacker is not required to leave anything as collateral. The cost of a crypto asset is manipulated in one transaction and then quickly resold in another transaction. The process used to conduct the attack is quick, and the hacker performs this several times before finishing and then disappears without leaving a trace [56]. A recent incident where this attack took place was on the *One Ring Space Protocol*. *One Ring Finance* had reported that attackers stole $1.4 million via a Flash Loan Attack from the blockchain platform. A total of $2 million was reported in losses after swap and flash loan fees were deducted. The attacker borrowed $80 million from *Solidly Flash Loans* to raise the price of LP tokens in the block spam. *OShare's* price was changed, and a large number of tokens were taken out of the protocol. However, the attack did not affect *OneRing (RING)* tokens, liquidity pools or farming opportunities in the *Fantom* space. The company began tracking how the attack occurred and realized that the contract was exploited using a self-destruct mechanism at a given block, making it impossible to determine the precise functions used to take the money. *Tornado Cash* was used to fund the attacker's Ethereum wallet, and the money was transferred into the protocol to obscure the transaction history. *One Ring Finance* took mitigating measures by restarting the vault, redeploying smart contracts, compensating users who lost money from the attack and improving security to address threats and vulnerabilities in the systems [57].

A *Parity Wallet Hack* is an attack where two transactions are sent to each affected contract. The first transaction is to obtain ownership of the wallet, and the second transaction is to move all the funds present in the wallet. An incident where the attack occurred resulted from a vulnerability to the *Parity MultiSig Wallet* version 1.5+. The attacker stole over 150,000 ETH (~30 million USD). The company reported that the vulnerability did not affect the second wallet *OpenZeppelin MultiSig Wallet*. The attack was the largest hack in terms of *ETH* stolen in the history of the *Ethereum* network. The attacker's account drained 153,037 ETH from three high-profile multi-signature contracts to store funds from past token sales. For the attack to be successful, the attacker sent two transactions to each affected contract: the first to obtain ownership of the MultiSig and the second to move all the funds from the wallet. The attacker used different functions to perform the hack. The first transaction was a call to '*initWallet*'. The function 'initWallet' was created to extract the wallet's constructor logic into a separate library. The contract forwards all unmatched function calls to the library using '*delegatecall*'. The next function, '*payable*', caused all public functions to be callable by any user, including 'initWallet', which changed the owners of the contract. The function 'initWallet' has no checks to prevent an attacker from calling it after a contract is initialized. The attacker exploited the no-check vulnerability and changed the '*m_owners*' state variable to a list containing the attacker's address. Only one confirmation was required to execute any transaction. The last step was implementing the execute function to transfer the money to an attacker-controlled account [58].

A *Distributed Autonomous Organization (DAO) Attack* took place when an attacker exploited a loophole in the *DAO's* smart contract. The *DAO Attack* came to be known as a *re-entrancy attack*. The fallback functions, which are special constructs in *Solidity* are exploited by the re-entrancy attack. The attacker deployed a smart contract that acted as an investor and deposited some *Ether* in the *DAO* [59]. The *DAO* was a decentralized autonomous organization launched on the Ethereum Blockchain and was hacked due to vulnerabilities in the code base. The tokens were locked up as the sale was to last for 28 days, after which the *DAO* would operate. Three weeks into the sale, the *DAO* had raised more than $150 million, but even before the sale could end, users expressed concerns about vulnerabilities in the *DAO's* code. An attacker exploited the vulnerability and stole funds from the *DAO* as programmers tried to fix the issue [60].

CONCLUSIONS AND RECOMMENDATIONS

Blockchain technology is one of the most revolutionary innovations of the modern age. It offers the promise of decentralized, secure and transparent systems that can change the way data is stored and managed; and how transactions are conducted. However, with this promise comes a set of security considerations that must be considered. The security of blockchain technology is an essential concern for anyone who wants to use it. Blockchain is a distributed database that is maintained by a network of nodes, which makes it highly secure against hacking attempts. Each node on the network has a copy of the database, which means that any changes to the data must be approved by most of the nodes. The change or modification is known as the consensus mechanism and ensures that any data stored on the blockchain is immutable and cannot be tampered with. However, blockchain technology is not completely secure. There are still several security issues that need to be considered when using the technology. Perhaps one of the most significant of these is the susceptibility of blockchains to 51% Attacks. A 51% Attack is when a single entity or group of entities controls more than 50% of the computing power on the network. With this level of control, the attackers can manipulate the blockchain, alter transactions and even steal funds. This is a significant concern for blockchain security and has been the subject of many debates and discussions in the blockchain community.

Wallet attacks have also been some of the most frequently occurring types of security breaches in the blockchain industry. Wallet attacks involve stealing a user's private keys or other credentials used to access their cryptocurrency wallet, giving the attacker access to the user's funds. Such attacks have resulted in significant losses for individuals and companies that rely on blockchain technology for financial transactions. As the blockchain industry continues to grow, it is essential to develop robust security measures to protect against these types of attacks.

To prevent 51% attacks, blockchain networks use various mechanisms such as PoW and PoS. PoW requires nodes to solve complex mathematical equations to validate transactions, while PoS uses a system of staking tokens to validate transactions. Both mechanisms are designed to make it difficult for any single entity to gain control of the network. Another significant security consideration in blockchains is the need for secure key management. Blockchain networks use public-key cryptography to secure transactions and protect user privacy. Public-key cryptography uses a pair of keys – one public and one private – to encrypt and decrypt data. However, the private key is the only way to access the funds stored in a blockchain wallet. If the private key is lost or stolen, the funds are lost forever. Therefore, it is essential to ensure that private keys are securely stored and protected. One way to protect private keys is by using a hardware wallet. A hardware wallet is a physical device that stores private keys offline, making it virtually impossible for hackers to steal them. Another way to protect private keys is by using a multi-signature wallet that requires multiple people to sign off on transactions.

Another very significant security consideration on blockchains is the potential for bugs or vulnerabilities in smart contracts. Smart contracts are self-executing contracts that are coded onto the blockchain. They allow for the automation of complex transactions and can be used to build decentralized applications (dApps). However, if there are bugs or vulnerabilities in a smart contract it can be exploited by attackers. This can result in the loss of funds or sensitive data. Therefore, it is essential to conduct regular security audits of smart contracts to ensure that they are free of bugs and vulnerabilities. In addition to these security considerations, there are several other security measures that should be taken when using blockchain technology. For example, it is essential to implement robust authentication and access control mechanisms to prevent unauthorized access to blockchain networks and wallets. It is also important to keep up to date with the latest security best practices in the blockchain industry.

In short, blockchain technology has the potential to transform various industries by providing secure and decentralized systems for data storage and transaction management. However, it is essential to be aware of the security considerations associated with blockchain and implement appropriate measures to mitigate the risks. This includes implementing robust authentication and access control mechanisms, conducting regular security audits, and keeping up to date with the latest security best practices in the blockchain industry. Tables 5.2 to 5.5 are aimed at helping readers with a list and explanations of various attacks for reference purposes.

Table 5.2 Summary of user-level attacks

Name of attack	Attack vector	Description	Mitigation strategy
Digital Signature Vulnerability	User	Based on asymmetric cryptography that uses both public and private keys. Digital signature is created when an algorithm yields the private and corresponding public keys.	Use Elliptic Curve Digital Signature Algorithm (ECDSA), which provides a higher level of security and is resistant to brute force attacks.
Hash Function Vulnerability	User	Attempt to find two strings of a hash function that produce the same hash result. A collision occurs when two pieces of data have the same digest.	Use a combination of different hash functions, known as hash function chaining or hash function iteration, to increase the level of security and reduce the risk of collision attacks.
Finney Attack	User	Occurs when an individual accepts an unconfirmed transaction on the network.	Implement confirmation times for transactions, delaying the execution of a transaction until it has been confirmed by multiple nodes.
Vector76 Attack	User	Takes advantage of minor bugs in the consensus system of Bitcoin. The aim of the attack is exchanges or exchange houses where attackers buy or sell cryptocurrency and tokens without being quickly discovered.	Implement improvements to the consensus algorithm of the blockchain. Improvements could include adding additional verification checks, modifying the transaction ordering process, or adjusting the block size and timing of block creation.
51% Attack	User	The attack occurs when a group of miners own more than 50% of the blockchain network's mining hash rate. Owning 50% gives the parties involved power to modify the blockchain and make changes to transactions.	The network's hash rate must be increased. This can be done by encouraging more miners to join the network, which would increase the amount of computational power required for an attacker to control 51% of the network's hash rate.
Phishing Attack	User	A cryptocurrency scam that involves deceiving users into disclosing private keys or personal information. The attack is done by sending a mass email or message to users.	Educate blockchain users on how to identify and avoid these attacks. This can include providing training on how to spot phishing emails, websites and other types of social engineering attacks. Users should be advised to never share their private keys, passwords or other sensitive information with anyone, and to always verify the authenticity of any blockchain-related communication before taking any action.

Table 5.3 Summary of network-level attacks

Name of attack	Attack vector	Description	Mitigation strategy
Side Chain Attack	Network	The attack is an attempt to extract secrets from a chip or a system. The extraction is achieved by measuring or analyzing physical parameters.	Measures such as encryption and secure communication protocols should be implemented to ensure the integrity and confidentiality of transactions. It's also important to regularly audit the side chain's code for vulnerabilities and to implement security patches as needed.
Sybil Attack	Network	The attack is arranged by assigning several identifiers to the same node. The attacker controls multiple nodes, and the user is surrounded by fake nodes that close all their transactions.	Implement a Proof-of-Work (POW) consensus algorithm. In PoW, nodes on the network compete to solve a computational puzzle, and the first node to solve it is rewarded with the ability to add a new block to the chain. This requires a significant amount of computing power, making it difficult for a malicious actor to create multiple Sybils and control the network.
Transaction Malleability Attack	Network	An attack that changes the unique transaction ID before it is approved or validated by the network. Attackers gain unauthorized access to the network and change transaction details.	Implement Segregated Witness (SegWit), which is a technology that separates transaction signatures from transaction data, making it more difficult for a malicious actor to modify the transaction ID.
Timejacking Attack	Network	The attack exploits a vulnerability in the Bitcoin timestamp handling. The attacker compels a node to accept an alternative blockchain by changing the node's network time.	Implement Network Time Protocol (NTP) time synchronization. This involves synchronizing the system clocks of all nodes on the network to a trusted time source, such as an atomic clock.
DDoS Attack	Network	A malicious attempt to disrupt traffic of a server, service or network by flooding the target with an excessive amount of traffic.	Implement DDoS protection measures such as rate limiting, traffic filtering and load balancing. This involves monitoring network traffic and filtering out malicious traffic before it can reach the network. Additionally, implementing load balancing can help distribute traffic across multiple servers.

Summary of blockchain-related attacks

Recommended best practices for organizations

As blockchain technology continues to expand, it is critical for organizations to prioritize security measures to protect their data and assets from cyber threats and attacks. By implementing best practices, organizations can mitigate risks and enhance the security of their blockchain infrastructure. Some key best practices are explored to ensure the security of

Table 5.4 Summary of system-level attacks

Name of attack	Attack vector	Description	Mitigation strategy
Selfish Mining Attack	System	Malicious miners withhold newly mined blocks and do not broadcast the blocks to the network. The blocks are released all at once, making other miners lose the originally published blocks.	Implement a dynamic difficulty adjustment algorithm. This algorithm adjusts the difficulty of mining based on the number of miners on the network. This makes it more difficult for a miner or a mining pool to control more than its fair share of block rewards.
Bribery Attack	System	The attacker publishes a transaction and waits until follow-up blocks have been published. A conflicting transaction is introduced as the attacker extends the branch of the first transaction. The funds are double spent, and the attacker earns a huge profit.	Implement multi-signature verification. This involves requiring multiple parties to sign off on transactions before they can be included in the blockchain. By requiring multiple parties to verify transactions, it becomes more difficult for a single miner or group of miners to include fraudulent transactions in the blockchain.
Crypto-jacking Attack	System	A device's computing power is abused with unauthorized access, and the computing power is used to mine cryptocurrency. The attack is delivered in the form of malicious software that infects devices used for mining.	Implement browser extensions that block mining scripts. These extensions can detect and block mining scripts that are embedded in websites, preventing them from using the victim's computing resources to mine cryptocurrency.
Malware Attack	System	Occurs when unauthorized actions are conducted using malware. The aim of the attack is to exfiltrate information, disrupt operations and demand payment.	Implement multi-factor authentication (MFA) for access to cryptocurrency wallets. MFA requires a user to provide more than one form of authentication to gain access to their wallet.

blockchain systems and networks. Based on the discussions in this chapter, the following best practices are recommended for organizations to mitigate risks caused by these attacks:

- *Information Systems Audit and Control Association (ISACA)* – Blockchain Preparation Audit Program

The program helps companies to manage the planning necessary to implement blockchain technology. Blockchain technology is the underlying network infrastructure that is connected to decentralized cryptocurrencies. The program assists auditors in identifying and developing key policies, procedures and controls to mitigate risks and streamline processes prior to a blockchain implementation.

The audit program concentrates on six key areas: Pre-implementation, Governance, Development, Security, Transactions and Consensus. The areas focus on the risks associated with using blockchain technology [61].

The program can help assess an organization's blockchain solution to determine whether it is adequately designed and operationally effective, and to identify blockchain risks that

Table 5.5 Summary of smart contract attacks

Name of attack	Attack vector	Description	Mitigation strategy
Flash Loan Attack	Smart Contract	Occurs when a platform's smart contract security is exploited. The attacker borrows a significant amount of money through a flash loan without leaving anything as collateral.	Implement threshold-based trading. This involves setting a threshold for the maximum amount of cryptocurrency that can be traded in a single transaction. By setting a threshold, the network can prevent large and sudden changes in the market.
Parity Wallet Hack	Smart Contract	Two transactions are sent to each affected contract. The first transaction is to obtain ownership of the wallet and the second transaction is to move the funds from the wallet.	Implement multi-signature verification for access to cryptocurrency wallets. This involves requiring multiple parties to sign off on transactions before they can be executed.
DAO (Decentralized Autonomous Organization) Attack	Smart Contract	A loophole in the DAO's smart contract is exploited. The fallback functions are exploited by the attack. The attacker deploys a smart contract that acts as an investor and deposits some Ether in the DAO.	Implement smart contract auditing. Smart contracts should be audited by experienced professionals to ensure that they are free from vulnerabilities and potential exploits.

could result in reputational and/or material impact. It can also provide organizations with a comprehensive perspective on blockchain technology with consideration for technical and non-technical factors [62]. For more information about the program, please visit https://store.isaca.org/s/store#/store/browse/detail/a2S4w000004KoDOEA0

- *Use Guardtime to Safeguard Data* [63].

This tool takes away the need to use keys for verification. If an attacker tries to alter the data, the system analyzes all chains, compares them to metadata packets and then excludes any that don't match. The only way to wipe the entire blockchain is to destroy every single separate node. If just one node remains running with correct data, the system can be restored if all other nodes are compromised. Guardtime's systems work in a way that the systems can detect when changes have been made to the data and constantly verify the changes. This helps to ensure that the blocks are not tampered with and that the data remains uncompromised.

For more information on Guardtime, please visit https://guardtime.com/

- *Conduct Blockchain Audits.*

Organizations should have secure technology that supports high-value transactions taking place on the blockchain. The process involves the use of sophisticated code analysis that identifies any loopholes in the system and eliminates any vulnerabilities in the system and applications [64]. Organizations can also do a blockchain code audit, which is a systematic and structured code review of blockchain implementation. The process is done manually and requires the use of static analysis tools. Security professionals should also keep in mind and take into consideration several steps when conducting an audit. The steps include [65]:

i. Defining the audit goal of the target system.

ii. Identifying components and associated data flows.

iii. Identifying security risks.

iv. Threat modeling.

v. Exploitation and remediation.

- The benefit of doing a *blockchain* audit is that it could lead organizations to create opportunities for the development of new services and the elimination of existing services, which can be replaced by technological systems [66].

In addition to the best practices suggested above, enterprises using blockchain systems should also:

- Consider emerging threats and new vulnerabilities.
- Use secure code frameworks and rigorous audits for creating secure code. This includes safe mathematical operations, authentication and authorization, or monetary transaction handling.
- Use static code analysis by eliminating bugs in code.
- Use dynamic code analysis – The *Maian* tool can be used to find vulnerable smart contracts. The tool is an example of hybrid static analysis with concrete execution discovering flaws missed by static analysis tools.
- Employ threat modeling to identify all components of the system, potential threats, vulnerabilities and controls provided by systems or platforms.
- Support governance policy creation by creating policies around consumer and asset protection. Developers also need to implement policies for patching live contracts and methods for handling incident response [67].

REFERENCES

[1] Tagade, K. (2022, August 6). An introduction to Blockchain Security. *Astra Security Blog*. Retrieved February 27, 2023, from https://www.getastra.com/blog/knowledge-base/blockchain-security/

[2] IBM. (n.d.). *What is Blockchain Security?* Retrieved February 27, 2023, from https://www.ibm.com/topics/blockchain-security#:~:text=Each%20new%20block%20connects%20to,transaction%20is%20true%20and%20correct

[3] Yatsenko, S. M. (2020, October 8). Blockchain attack vectors: Vulnerabilities of the most secure technology. *Apriorit*. Retrieved October 6, 2022, from https://www.apriorit.com/dev-blog/578-blockchain-attack-vectors

[4] OWASP Top Ten 2017. (n.d.). *2017 top 10. | OWASP Foundation*. Retrieved March 10, 2023, from https://owasp.org/www-project-top-ten/2017/Top_10

[5] OWASP Foundation. (n.d.). *Owasp Top Ten*. Retrieved March 10, 2023, from https://owasp.org/www-project-top-ten/

[6] OWASP Top 10:2021. (n.d.-a) *A01:2021 – Broken Access Control*. Retrieved March 10, 2023, from https://owasp.org/Top10/A01_2021-Broken_Access_Control/

[7] OWASP Top 10:2021. (n.d.-b). *A02:2021 – Cryptographic Failures*. A02 Cryptographic Failures. Retrieved March 10, 2023, from https://owasp.org/Top10/A02_2021-Cryptographic_Failures/

[8] OWASP Top 10:2021. (n.d.-c). *A03:2021 – injection*. A03 Injection. Retrieved March 10, 2023, from https://owasp.org/Top10/A03_2021-Injection/

[9] OWASP Top 10:2021. (n.d.-d). *A04:2021 – insecure design*. A04 Insecure Design. Retrieved March 10, 2023, from https://owasp.org/Top10/A04_2021-Insecure_Design/

[10] OWASP Top 10:2021. (n.d.-e). *A05:2021 – security misconfiguration*. A05 Security Misconfiguration - Retrieved March 10, 2023, from https://owasp.org/Top10/A05_2021-Security_Misconfiguration/

[11] OWASP Top 10:2021. (n.d.-f). *A06:2021 – vulnerable and outdated components*. A06 Vulnerable and Outdated Components. Retrieved March 10, 2023, from https://owasp.org/Top10/A06_2021-Vulnerable_and_Outdated_Components/

[12] OWASP Top 10:2021. (n.d.-g). *A07:2021 – identification and authentication failures*. A07 Identification and Authentication Failures. Retrieved March 10, 2023, from https://owasp.org/Top10/A07_2021-Identification_and_Authentication_Failures/

[13] OWASP Top 10:2021. (n.d.-h). *A08:2021 – software and Data Integrity Failures*. A08 Software and Data Integrity Failures. Retrieved March 10, 2023, from https://owasp.org/Top10/A08_2021-Software_and_Data_Integrity_Failures/

[14] OWASP Top 10:2021. (n.d.-i). *A09:2021 – security logging and monitoring failures*. A09 Security Logging and Monitoring Failures. Retrieved March 10, 2023, from https://owasp.org/Top10/A09_2021-Security_Logging_and_Monitoring_Failures/

[15] OWASP Top 10:2021. (n.d.-j). *A10:2021 – server-side request forgery (SSRF)*. A10 Server Side Request Forgery (SSRF). Retrieved March 10, 2023, from https://owasp.org/Top10/A10_2021-Server-Side_Request_Forgery_%28SSRF%29/

[16] Blufolio. (n.d.). *Security of Blockchain*. Retrieved October 12, 2022, from https://blufol.io/security-of-blockchain/

[17] Ifegwu, O. (n.d.). Finality. *Binance Academy*. Retrieved March 4, 2023, from https://academy.binance.com/en/glossary/finality

[18] Tunggal, A. T. (2023, January 8). *What is an attack vector? 16 common attack vectors in 2023*. UpGuard. Retrieved March 13, 2023, from https://www.upguard.com/blog/attack-vector

[19] Aakurathi, S. K. (2020, June 27). *Different types of attacks on a system*. Medium. Retrieved October 8, 2022, from https://srujan-aakurathi.medium.com/different-types-of-attacks-on-a-system-e99618ada736#:~:text=In%20Operating%20Systems%20attacks%2C%20%E2%80%9Cattackers,services%20and%20ports%20by%20default.

[20] Cynet. (2022, August 7). *Network attacks and network security threats*. Retrieved October 9, 2022, from https://www.cynet.com/network-attacks/network-attacks-and-network-security-threats/#:~:text=What%20Is%20a%20Network%20Attack%3F,or%20perform%20other%20malicious%20activity.

[21] Synopsys. (n.d.). *What is Blockchain and how does it work?* Retrieved October 8, 2022, from https://www.synopsys.com/glossary/what-is-blockchain.html

[22] Frankenfield, J. (2022a, March 24). *What are smart contracts on the Blockchain and how they work?* Investopedia. Retrieved October 9, 2022, from https://www.investopedia.com/terms/s/smart-contracts.asp

[23] Salehi, F. (n.d.). *Vulnerabilities of digital signatures and protection methods*. ShaKeyLead. Retrieved October 12, 2022, from https://emza.me/en/vulnerabilities-of-digital-signatures-and-protection-methods/

[24] Sheridan, K. (2022, January 5). New attack campaign exploits Microsoft Signature Verification. *Dark Reading*. Retrieved December 3, 2022, from https://www.darkreading.com/attacks-breaches/new-attack-campaign-exploits-microsoft-signature-verification

[25] Arampatzis, A. (2020, April 6). What is the hashing function and can it become vulnerable? *Venafi*. Retrieved December 3, 2022, from https://www.venafi.com/blog/what-hashing-function-and-can-it-become-vulnerable#:~:text=A%20hash%20function%20attack%20is,same%20digest%20as%20shown%20above.

[26] Bit2Me Academy. (n.d.-a). *What is a Finney hack or Finney attack?* Retrieved October 13, 2022, from https://academy.bit2me.com/en/que-es-un-hackeo-finney-ataque-finney/

[27] Bit2Me Academy. (n.d.-b). *What is vector attack 76?* Retrieved October 13, 2022, from https://academy.bit2me.com/en/que-es-ataque-de-vector-76/

[28] Blockchain attacks: Vector attack 76. (2021, April 20). *spacebot.com*. Retrieved December 3, 2022, from https://spacebot.group/blockchain/blockchain-attacks-vector-attack-76/

[29] Frankenfield, J. (2022, September 28). *51% attack: Definition, who is at risk, example, and cost*. Investopedia. Retrieved October 12, 2022, from https://www.investopedia.com/terms/1/51-attack.asp

[30] Hill, E. (2020, January 27). *Bitcoin Gold suffers 51% attack with $72,000 stolen*. Yahoo! News. Retrieved December 5, 2022, from https://ca.news.yahoo.com/bitcoin-gold-suffers-51-attack-140039732.html?guccounter=1&guce_referrer=aHR0cHM6Ly93d3cuZ29vZ2xlLmNvbNvb8&guce_referrer_sig=AQAAAGb8E1BsTwD9iutTHx6L9DtjQiRzFaNjZyYQztcbWk-2VivnaJNmik0GZUlm3y6XBEO64auO1z3S6ZJPXtMZjwmAw36RWtkOrjUx7hY84TskjSAuSdNkLU82gA7EawmlTah6Fyl7izJZGmnNuFbE-k1ljmo7oNQsz3T3ujvG8NV2

[31] Cointelegraph. (2022, July 27). *What is a phishing attack in crypto, and how to prevent it?* Cointelegraph. Retrieved March 20, 2023, from https://cointelegraph.com/blockchain-for-beginners/what-is-a-phishing-attack-in-crypto-and-how-to-prevent-it

[32] Roth, E. (2022, August 8). *Twilio suffers data breach after its employees were targeted by a phishing campaign.* The Verge. Retrieved March 20, 2023, from https://www.theverge.com/2022/8/8/23296923/twilio-data-breach-phishing-campaign-employees-targeted

[33] Bhunia, S., & Tehranipoor, M. (2019). Side channel attack. Side Channel Attack - an overview | ScienceDirect Topics. Retrieved December 5, 2022, from https://www.sciencedirect.com/topics/computer-science/side-channel-attack

[34] Faife, C. (2022, October 7). Hacker steals $570 million of crypto tokens from Binance's BNB Chain. *The Verge.* Retrieved November 14, 2022, from https://www.theverge.com/2022/10/7/23392424/hacker-steals-570-million-binance-bnb-chain

[35] Buford, J. F., & Lua, E. K. (2009). Sybil attack. Sybil Attack - an overview | ScienceDirect Topics. Retrieved December 5, 2022, from https://www.sciencedirect.com/topics/computer-science/sybil-attack

[36] Imperva. (n.d.). *What is a Sybil attack: Examples & prevention: Imperva.* Retrieved March 5, 2023, from https://www.imperva.com/learn/application-security/sybil-attack/#:~:text=A%20Sybil%20attack%20uses%20a,of%20influence%20in%20the%20network.

[37] Caffyn, G. (2015, March 14). Chainalysis CEO denies 'Sybil attack' on Bitcoin's network. *CoinDesk Latest Headlines RSS.* Retrieved December 5, 2022, from https://www.coindesk.com/markets/2015/03/14/chainalysis-ceo-denies-sybil-attack-on-bitcoins-network/

[38] Poston, H. (2020, October 13). Attacking the blockchain network. *Infosec Resources.* Retrieved October 9, 2022, from https://resources.infosecinstitute.com/topic/attacking-the-network/#:~:text=Blockchains%20are%20vulnerable%20to%20network,the%20only%20possible%20attack%20vector

[39] Mutuku, K. (2022, November 30). Transaction malleability: A legacy blockchain vulnerability. *Web 3 Africa.* Retrieved December 8, 2022, from https://web3africa.news/2022/11/30/news/transaction-malleability-in-blockchain-vulnerability/

[40] Norry, A. (2020, March 31). *The history of the Mt Gox hack: Bitcoin's biggest heist.* Blockonomi. Retrieved March 5, 2023, from https://blockonomi.com/mt-gox-hack/

[41] NFTing. (2022, October 25). *What is timejacking.* Medium. Retrieved March 19, 2023, from https://nfting.medium.com/timewhat-is-timejacking-5ab8f08ae82f

[42] Cloudfare. (n.d.). *What is a distributed denial-of-service (ddos) attack?* Retrieved March 19, 2023, from https://www.cloudflare.com/learning/ddos/what-is-a-ddos-attack/

[43] Palo Alto Networks. (n.d.). *What is a denial of service attack (DoS)?* Retrieved March 19, 2023, from https://www.paloaltonetworks.com/cyberpedia/what-is-a-denial-of-service-attack-dos#:~:text=A%20Denial%2Dof%2DService%20(,information%20that%20triggers%20a%1120crash.

[44] Rooney, K. (2018, June 5). *Cryptocurrency exchange Bitfinex briefly halts trading after cyberattack.* CNBC. Retrieved March 19, 2023, from https://www.cnbc.com/2018/06/05/cryptocurrency-exchange-bitfinex-briefly-halts-trading-after-cyber-attack.html

[45] Zhou, C., Xing, L., Guo, J., & Liu, Q. (2022). *Bitcoin selfish mining modeling and dependability analysis* - IJMEMS. Retrieved December 7, 2022, from https://ijmems.in/cms/storage/app/public/uploads/volumes/2-IJMEMS-21-0720-7-1-16-27-2022.pdf

[46] Gutteridge, D. (2018, May 21). *Japanese cryptocurrency Monacoin hit by selfish mining attack.* Yahoo! Finance. Retrieved March 6, 2023, from https://ca.finance.yahoo.com/news/japanese-cryptocurrency-monacoin-hit-selfish-205031219.html

[47] Bonneau, J. (n.d.). Why buy when you can rent? – jbonneau.com. Retrieved December 8, 2022, from https://jbonneau.com/doc/BFGKN14-bitcoin_bribery.pdf

[48] Sandor, K. (2022, March 22). What is cryptojacking? how to protect yourself against Crypto Mining Malware. *CoinDesk Latest Headlines RSS.* Retrieved December 5, 2022, from https://www.coindesk.com/learn/what-is-cryptojacking-how-to-protect-yourself-against-crypto-mining-malware/

[49] Browne, R. (2018, February 21). *Hackers hijack Tesla's cloud system to mine cryptocurrency.* CNBC. Retrieved December 5, 2022, from https://www.cnbc.com/2018/02/21/hackers-hijack-teslas-cloud-system-to-mine-cryptocurrency-redlock.html

[50] Colyer, A. (2017, December 7). *Be selfish and avoid dilemmas: fork-after-withholding attacks on Bitcoin*. The morning paper. Retrieved March 11, 2023, from https://blog.acolyer.org/2017/12/07/be-selfish-and-avoid-dilemmas-fork-after-withholding-attacks-on-bitcoin/

[51] Zhang, Y., Chen, Y., Miao, K., Ren, T., & Yang, C. (2022, November 24). A novel data-driven evaluation framework for fork after withholding attack in blockchain systems. *ResearchGate*. Retrieved December 8, 2022, from https://www.researchgate.net/publication/365741305_A_Novel_Data-Driven_Evaluation_Framework_for_Fork_after_Withholding_Attack_in_Blockchain_Systems

[52] Browne, R. (2022a, August 2). *Hackers drain nearly $200 million from crypto startup in 'free-for-all' attack*. CNBC. Retrieved March 11, 2023, from https://www.cnbc.com/2022/08/02/hackers-drain-nearly-200-million-from-crypto-startup-nomad.html

[53] Rapid7. (n.d.). *What is a malware attack? Definition & Best Practices*. Retrieved October 12, 2022, from https://www.rapid7.com/fundamentals/malware-attacks/

[54] Clop ransomware. (n.d.). *Mimecast*. Retrieved December 5, 2022, from https://www.mimecast.com/content/clop-ransomware/

[55] Werapun, W., Karode, T., Arpornthip, T., Suaboot, J., Sangiamkul, E., & Boonrat, P. (2022, December 30). The Flash Loan Attack Analysis (FAA) framework-A case study of the Warp Finance Exploitation. *MDPI*. Retrieved February 27, 2023, from https://www.mdpi.com/2227-9709/10/1/3#:~:text=The%20flash%20loan%20attack%20is,certain%20assets%20on%20the%20platform.

[56] Bybit Learn. (2021, December 5). What is a flash loan attack - and how do I prevent it? *Bybit Learn*. Retrieved October 12, 2022, from https://learn.bybit.com/defi/what-is-a-flash-loan-attack/

[57] Bannister, A. (2022, March 24). *Flash loan attack on One ring protocol nets crypto-thief $1.4 million*. The Daily Swig | Cybersecurity news and views. Retrieved December 5, 2022, from https://portswigger.net/daily-swig/flash-loan-attack-on-one-ring-protocol-nets-crypto-thief-1-4-million

[58] Palladino, S. (2017, July 19). *The parity wallet Hack explained*. OpenZeppelin blog. Retrieved October 13, 2022, from https://blog.openzeppelin.com/on-the-parity-wallet-multisig-hack-405a8c12e8f7/

[59] Jones, W. (2022, October 13). *The Dao attack: Understanding what happened*. crypto.news. Retrieved December 8, 2022, from https://crypto.news/learn/the-dao-attack-understanding-what-happened/

[60] Cryptopedia. (2022, March 16). *What was the DAO?* Gemini. Retrieved October 26, 2022, from https://www.gemini.com/cryptopedia/the-dao-hack-makerdao

[61] ISACA. (2019, March 19). *ISACA issues new blockchain CASB Solutions and GDPR audit programs*. Retrieved March 20, 2023, from https://www.isaca.org/why-isaca/about-us/newsroom/press-releases/2019/isaca-issues-new-blockchain-casb-solutions-and-gdpr-audit-programs#:~:text=Blockchain%20Preparation%20Audit%20Program%20is,US%20%2449%20for%20non%2Dmembers.

[62] Ebenezer, V. (n.d.). *Mitigate risks when deploying blockchain technology*. Infosecurity Magazine. Retrieved October 9, 2022, from https://www.infosecurity-magazine.com/infosec/deploying-blockchain-technology-1-1-1/

[63] Horbenko, Y. (n.d.). Using blockchain technology to boost cyber security. *SteeKiwi*. Retrieved November 17, 2022, from https://steelkiwi.com/blog/using-blockchain-technology-to-boost-cybersecurity/

[64] Gondek, C. (n.d.). *What is a blockchain security audit?* OriginStamp. Retrieved October 5, 2022, from https://originstamp.com/blog/what-is-a-blockchain-security-audit/

[65] Keshri, A. (2022, August 22). How to perform a blockchain security audit? *Astra Security Blog*. Retrieved December 8, 2022, from https://www.getastra.com/blog/security-audit/blockchain-security-audit/

[66] Elommal, N., & Manita, R. (2022). How blockchain innovation could affect the audit profession: a qualitative study. *Cairn Matieres A Reflexion*. Retrieved December 8, 2022, from https://www.cairn.info/revue-journal-of-innovation-economics-2022-1-page-37.htm

[67] Riedesel, S. (2018, March 7). *How can blockchain applications adapt and adopt software security best practices?* Application Security Blog. Retrieved October 9, 2022, from https://www.synopsys.com/blogs/software-security/blockchain-security-best-practices/

Chapter 6

Digital assets valuation and financial reporting

Ankita Vashisth, Kolawole Salako and Pramitha Pinto

BACKGROUND

Blockchain is a decentralized database or electronic ledger used to securely store ownership records of digital assets. According to *Vincent*, digital assets are digital data files that can be owned and transferred by individuals and used as a means of storing intangible contents like computerized artworks, videos and contract documents, or as a currency for making transactions. Cryptocurrencies, asset-backed stablecoins and non-fungible tokens (NFTs) are examples of digital assets [1].

A cryptocurrency is a digital asset created with the intention of serving as a medium of exchange that uses cryptography to safeguard transactions, regulate the creation of new value units and confirm the transfer of assets [2]. They are decentralized in nature and executed on distributed networks without any central authority [3]. Cryptocurrency uses a distributed ledger or blockchain technology to facilitate secure transactions. As such, it employs a peer-to-peer digital network that uses an open-source decentralized autonomous organization (DAO) [2]. According to *AsianMarketCap official* report, the irreversible and irrevocable nature of cryptocurrencies ensures that transactions recorded on the blockchain cannot be reversed. Cryptocurrency transactions are not directly associated with an identity, giving users some degree of anonymity. Due to the underlying coding mechanism, cryptocurrencies can be designed to have an immutably finite quantity [3].

Benefits of cryptocurrencies

As reported by *Bunjaku et al*, the integrity of the entire system is guaranteed because the payments made using a cryptocurrency-based system cannot be reversed, replicated, counterfeited or consumed more than once [4]. Limitless transaction exchange is another benefit as any cryptocurrency wallet holder can pay another wallet holder irrespective of the location because the transaction cannot be stopped or controlled. Cryptocurrencies provide a cost-effective operational approach by eliminating intermediary parties' commissions. The speed of payment or transactions is quite high due to complex computational involvement [4]. Additionally, decentralized cryptocurrencies typically make secure payment options because they are based on cryptography and blockchain security. *Nibley* considered cryptocurrencies as an inclusive financial system, where people who don't have access to the traditional financial system can take advantage of cryptocurrency to participate outside of the system [5].

DOI: 10.1201/9781003462033-7

Limitations of cryptocurrencies

Cryptocurrencies are very volatile as government policies and regulations can cause instability in the system. According to *Bunjaku et al* report, there is no central issuer for the currency or official legal institution to provide guarantees in the event of bankruptcy [4]. Moreover, cryptocurrency is not backed by banks or other financial institutions, which can guarantee liabilities and pay off outstanding balances with other assets [6]. The lack of institutional backing and hacking risks makes the future of cryptocurrencies unfavorable. As such, cryptocurrencies can be used to launder money or finance illegal operations [4]. In addition, the cryptocurrency acceptance rate is low because cryptocurrencies are still relatively new and only a small number of crypto-ATMs exist [6].

Several countries are yet to put in place regulatory frameworks to adopt cryptocurrencies. For example, in India, virtual currencies like Bitcoin are not accepted as legal money [7]. Another drawback identified is the legal acceptability and regulations of cryptocurrencies across the globe. Cryptocurrencies are unlikely to become stable enough to be used as money in their current state. Cryptocurrency valuation is difficult due to a lack of comparable trades, price differences between buy and sell orders, and the exchange used for the trade [8].

Attributes of non-fungible tokens

NFTs are non-monetary digital assets that are often developed using the same programming that supports cryptocurrencies, which cannot be traded or swapped for one another, unlike other crypto assets [9]. An NFT is a type of crypto asset that functions on a blockchain and assigns a unique identification to each gathered object. As the identification cannot be replicated, the owner of the digital object is established [9]. Due to the distinctive data of NFTs, it is simple to confirm ownership and transfer tokens between owners [9]. The ability to publicly verify the associated token metadata best describes NFT's qualities [10]. As such, NFTs are exclusive assets that can only be owned by one person [9]. Buying, selling and minting NFTs are open to the public in a transparent manner. Availability is established as issued NFTs and tokens are accessible for sale and purchase because of no system downtime. NFTs are also tamper-resistant as data cannot be manipulated and they can be traded and exchanged freely [10]. Additionally, NFTs have been designed to be indivisible in order to maximize utilization. One of the fundamental ideas underlying the value of NFTs is scarcity. NFT producers may therefore manufacture a significant number of NFTs, but with a cap to ensure scarcity [11]. In short, NFTs are tokens that establish ownership of an underlying asset in a cryptographically secure manner. In addition, NFTs promote inclusive growth as they support equitable growth as NFTs link content producers from various sectors together in one ecosystem, leading to growth that is fair for all participants [12].

Limitation of NFTs

Currently, the value of an NFT is solely dependent on its visual and sentimental appeal. Since it is impossible to estimate its value as a long-term investment, it is merely a highly speculative asset at this time. There have been a few NFT security breaches, primarily by hackers, as many exchanges have inefficient or out-of-date security mechanisms in place. Blockchain transactions, in addition to the creation and sale of NFTs, consume a significant amount of mining power. A growing market for NFTs, according to some scientists, could further harm the environment that is rapidly depleting [13]. *Lidén* reported that NFT's current consensus method (proof-of-work) consumes a large amount of energy and causes negative environmental impact [14]. The distribution or duplication of the original NFT across platforms cannot be controlled by the ownership of the original NFT [13]. Due to the immutable nature of NFTs, there are some concerns when it comes to data protection regulations like the GDPR (General Data Protection Regulation) in the EU. It is challenging to comply with the "right to be forgotten" because it is impossible to delete potential user data saved on the blockchain [14]. Furthermore, legally enforcing smart contracts through ownership verification can be challenging due to the anonymity of NFTs and the need to adhere to existing legal contract frameworks across various jurisdictions [14].

NFT valuation and financial reporting

Intellectual property concerns are the next key issue on the list of NFT dangers and challenges. As a result, when an NFT is acquired, the owner acquires simply the right to utilize it, rather than intellectual property rights. As reported by *Rehman et al (2021)*, cyber security and fraud risks are increasing as the digital world expands and NFT transactions increase. The mention of smart contract risk and challenges of NFT maintenance is also one of the prominent concerns in the NFT landscape presently. Smart contract development and security are other critical concerns in the NFT environment. Along with that, smart contract transactions and code are both immutable, which signifies that programmers are responsible for ensuring the security of the code and transactions cannot be altered. There are no defined standard procedures for designing smart contracts that developers should follow across the project [15]. According to *101 Blockchains*, the majority of NFTs that are offered on the market are sold as securities. NFTs, however, are now included in the Supreme Court's definition of an investment contract. For NFTs to demonstrate their eligibility as securities, they must adhere to the Howey Test's unique requirements. Moreover, AML (anti-money laundering) and CFT (combating the financing of terrorism) issues may arise as a result of the decentralized nature of NFT and without any central authority monitoring and jurisdictional challenges [16].

Similarities and differences between NFT and cryptocurrencies

As mentioned, the similarity between cryptocurrencies and NFTs is that both are digital assets that reside on blockchain and can be bought and sold. The blockchain keeps track of all transactions, making it simple to confirm the legitimacy of an NFT or coin. The fundamental distinction is that the value of cryptocurrency is exclusively determined by its economic function as a currency or investment. NFTs, on the other hand, have economic and non-economic value. For example, artists may use NFTs to share, market and even sign their works, which an investor or collector can then purchase using Bitcoin. The *fungibility* aspect is the most significant distinction, where crypto coins are fungible or mutually interchangeable and NFTs have a unique value. The manner in which crypto coins are put into circulation is another distinction, as NFTs are minted while coins are mined [17].

Top ten cryptocurrencies and NFTs

According to the *Coin Market cap* exchange as of November 2022, the total market capitalization of cryptocurrencies and NFTs was $845,176,034,194 and $14,079,274,931 [18]. The details of the top ten cryptocurrencies and NFTs' market capitalization are tabulated in Table 6.1.

Table 6.1 Top ten cryptocurrencies and NFTs

S/n	Cryptocurrencies	Market cap	NFTs	Market cap
1	Bitcoin (BTC)	$319,330,396,284	Flow	$1,288,602,357
2	Ethereum (ETH)	$152,329,553,052	Chiliz	$1,273,271,103
3	Tether	$68,268,232,163	Tezos	$956,114,177
4	BNB	$45,861,824,585	Apecoin	$940,884,619
5	USD coin	$43,994,026,853	Theta Network	$933,925,639
6	Binance USD	$23,004,665,124	The Sandbox	$923,177,057
7	XRP	$18,667,521,700	Decentraland	$896,756,922
8	Cardanao ADA	$11,987,473,142	Axie Infinity	$652,201,497
9	Dogecoin	$10,869,981,560	PancakeSwap	$599,664,621
10	Polygon	$8,820,369,104	Enjincoin	$339,349,078

Source: https://coinmarketcap.com/

DISCUSSION

International Financial Reporting Standards (IFRS) IAS 38 specifies the standards for identifying and valuing intangible assets, as well as the information that must be disclosed. Any monetary or non-financial item that cannot be physically handled is considered an intangible asset [19]. In a similar manner, the *International Financial Reporting Interpretations Committee* (IFRIC) notes that virtual currencies satisfy IFRS IAS 38's definition of an intangible asset because they can be sold or transferred separately from the holder and do not grant the holder the right to receive a predetermined or fixed amount of cash [20]. A separable asset or one that is derived from contractual or other legal rights makes it an identifiable non-monetary asset [19]. IFRIC contends that while some virtual currencies can be used to pay for goods and services, none can be regarded as cash or as a medium of measurement that would be the basis for all accounting calculations [20]. The cost of developing an intangible asset internally is often not easy to differentiate from the cost of improving the entity's operations. The costs of producing other internally generated intangible assets are categorized according to whether they occur during the development or research phases. Intangible assets are first valued at cost. An organization typically measures an intangible asset at cost less accumulated amortization after initial recognition. In rare instances where fair value may be established by reference to an active market, the entity may decide to measure the asset at fair value. An intangible asset with a limited useful life is not depreciated but amortized and undergoes an annual impairment test. The gain or loss on the sale of an intangible asset is counted toward profit or loss [19].

In 2014, the Canadian Senate reviewed the issue of virtual currency taxation and the *Canada Revenue Agency* (CRA) recommended actions to help Canadians understand how to comply with digital assets taxes. In order to file tax returns, individuals need to know how to value their cryptocurrencies, which relies on whether it is a capital property or inventory. If cryptocurrency is held as a capital asset, an adjusted cost basis must be recorded and tracked so that capital gains can be accurately reported. If cryptocurrency is held as an inventory, either value each record in the inventory at its cost when obtained or value the entire inventory at its fair market value at year-end [21]. When it comes to applying standard methods of valuation to crypto assets, there is a healthy amount of skepticism. This has increased due to the tremendous volatility of Bitcoin and other cryptocurrencies. *Ernest Young* (EY) report states that it is still appropriate and applicable to determine these assets' intrinsic value using conventional valuation procedures and principles. As market enthusiasm drive pricing, such strategies should take into account the intrinsic volatility or riskiness of these assets. Financial reporting, taxation and investment concerns are the three main factors driving the demand for a thorough approach to crypto-asset valuation [22].

Types of valuation methodology

The challenge of crypto asset valuations increases with the creation of new asset classes. It is crucial to understand the difference between an asset's market value, which is represented by the market price, and its intrinsic value. Often, psychological considerations influence market prices as such investors rely their judgments more on what other market participants are doing than on thorough research [23].

One of the first considerations to examine when valuing digital assets is the basis of valuation, that is, if the value is for transaction purposes, taxation, financial reporting or some other reason. For financial reporting purposes, it must be determined if the digital asset is cash or cash equivalent, a financial instrument, an intangible asset or an inventory. The valuation foundation and classification in the appropriate framework both have an influence on the eventual method of establishing value. According to the *Financial Accounting Standards Board (FASB)* and the *International Accounting Standards Board (IASB)*, many digital assets are similar to indefinite-lived intangible assets or even inventories. Depending on the accounting system employed, this may have an influence on how they are measured and reported. For example, investment firms that hold digital assets must measure and report their investments at fair value regardless of how they are categorized [24]. Moreover, one of the key elements influencing market pricing is the favorable media publicity. Financial reports assist in determining a stock's worth in traditional financial analysis of equities. This is no longer valid for crypto assets, and in order to determine their monetary worth, new valuation methodologies must be created [23].

According to IFRS 13:62, the three widely used valuation methods that are in line with International Accounting Standards are the market approach, cost approach and income approach, which are discussed below [25].

Market Approach: The process of assigning a value to a company based on market forces in comparable circumstances is known as the market value approach to business valuation [25]. The market technique evaluates the enterprise value of a subject firm based on the cash flow and profitability, consistent growth profile and company continuing forever of comparable companies [26]. Technical core, token model, underlying value, valuation trajectory, user experience, ecosystem breadth, consensus protocol and governance are factors influencing the adoption, success and price of digital assets that are different from traditional assets [26]. The major benefit of the market value technique is that it is based on publicly available data on comparable transactions. The market technique evaluates the worth of comparable NFTs, assuming that comparable assets would sell at comparable prices [27]. The value of an NFT is determined by the amount the next buyer will pay, not by what the previous buyer paid [25].

Cost Approach: The average variable cost of production establishes a lower bound on miners' assessed worth of the digital asset. If mining cryptocurrency marketplaces are robust and moderately competitive, the relative difficulty of mining such coins might be seen as a plausible signal of their relative worth [24]. This approach of cryptocurrency assumes that a person can choose to either mine coins for a predetermined number of units as a reward or buy on an exchange. Mining costs include both hardware and electricity costs to operate the hardware [28].

Income Approach: Converts future funds to a single current amount and reflects current market expectations about those future funds [25]. According to *Larsen and Weber*, the value of a digital asset with security-like properties is determined by discounting the cash flows connected with the asset across time [24]. Additional valuation methods that can be applicable to digital assets are the following:

Asset Approach: The fair value of a company's underlying assets less its outstanding liabilities is the basis for the asset approach's valuation. In contrast to the market and *Discounted Cash Flow (DCF)* approaches, which are time-adjusted, the asset approach takes into account future returns that are greater than net assets and focuses on liquidation-adjusted net assets [26].

Mark-to-Market Methodology: The method is a term that involves adjusting an asset's value to reflect its current market value. The method can either be used to measure the fair value of funds that are subject to periodic fluctuations, or it can be used in financial markets to represent the current and fair market value of investments. In the first method, the value of the assets is kept at the original purchase cost while in the latter approach asset values are based on amounts that can be exchanged under market conditions [29].

The *Internal Revenue Code (IRC) of 1986* is the federal statutory tax legislation in the United States. *IRC Section 475* requires the application of mark-to-market rules to securities held by dealers. A security dealer is either a taxpayer who regularly buys or sells securities to a customer in the ordinary course of trade or one who offers to solicit trading securities with customers in the ordinary course of trade. According to IRC Section 475, cryptocurrencies like Bitcoin and Ethereum do not appear to meet the definition of security therefore, it is unlikely that the mark-to-market rule will apply. On the other side, if a cryptocurrency is an equity token (a type of security token that symbolizes the ownership interest in an underlying asset), then the concept of *mark-to-market* is subject to IRC Section 475 and this cryptocurrency will be marked as a *Fair Market Value* (FMV) at the end of the year [30].

Corporate financial reporting

Blockchains have the capacity to improve the quality of information that reaches investors in two different ways: by making the accounting data more reliable and timely. Regarding trust, if businesses keep their financial records on blockchains, the chances of accounting-related fraud and manipulation may decrease significantly. In addition, transactions between firms would also be much more transparent. Regarding timing, real-time updating of accounting data would be possible since blockchain-based bookkeeping would make every transaction in a company's ledger instantly accessible. In addition, this information would also be instantly made available to regulators as well as insiders within the company [31].

Blockchain and real-time accounting

The accounting information system makes it possible for financial data to be distributed in a methodical manner throughout the business in order to give users access to the data whenever and wherever needed. As such, blockchain technology has emerged as a promising technological solution for accounting information systems. It alters accounting theory in significant ways, and as the use of digital currencies by businesses grows, its effects on reporting become apparent [32]. Since blockchains are basically immutable ledgers, its information cannot be altered or deleted. As a result, it can prove very valuable as a reliable and up-to-date ledger for a company's accounting records. Financial statements are summaries of a company's ledger activity over a specific time period that are prepared on a regular basis. The financial statements' accuracy is then evaluated by an auditor. External parties need to trust the auditor's report and that the organization has not presented any false data to the auditor. As such, blockchain technology has the potential to play a crucial role in the area of trust and reliability. A company's entire ledger would be instantly visible and accessible in real time, if all of its business transactions were stored on a blockchain with a permanent time stamp for each transaction. A blockchain may be able to perform many of the tasks that an auditor performs in the accounting industry in a manner that is significantly more effective and timely. For investors, blockchain technology could enhance the quality of accounting information by making the information more reliable and accurate [31].

Integration into accounting processes and systems

The use of cryptocurrencies and other digital assets is increasing among consumers, businesses and governments. In spite of this increased attention, the financial reporting for these digital assets does not fit perfectly with the *Generally Accepted Accounting Principles (GAAP)* guidelines that exist in the United States [33]. As of this writing, there are no set guidelines on how to account for cryptocurrencies [34]. Certified Public Accountants (CPAs) and accounting firms have requested the FASB to address this rising concern and to consider publishing updated guidelines in line with these new assets, but it is unclear if the FASB will consider these assets on their agenda. Crypto reporting issues with current accounting standards:

Changes in Value: Crypto assets cannot be accounted for using the same standards that apply to cash because virtual currencies aren't legal tender – (unless you are in El Salvador) – how digital assets will be handled from a regulatory standpoint hasn't been confirmed by the government.

Intangible Asset: Under GAAP or IFRS, companies usually accept cryptocurrencies on the balance sheet at their cost basis, and as cryptocurrencies are impaired they should be recorded as a loss rather than as an indefinite-lived intangible asset.

Recorded Losses, not Gains: Only unrealized losses are recorded in the United States [33].

Adoption of cryptocurrency and NFTs in financial statements

According to the FASB, businesses should track crypto assets using fair value accounting, recording gains and losses in current period comprehensive income, but this decision is not yet final. This will be a significant advancement since it brings us one step closer to the time when it will be feasible for businesses to include crypto assets in their balance sheets. The low adoption of crypto assets by businesses is frequently attributed to the lack of clarity in accounting standards. As more of the transactions move to blockchain, a significant portion of the accounting industry will be replaced [34]. The FASB Board adopted some decisions concerning the measurement of in-scope crypto assets: crypto assets should be assessed at fair value, and changes in fair value should be recognized in the current period comprehensive income. Alternative measurements such as historical cost less impairment for crypto assets that are not traded in an active market are not allowed. Fees, commissions and other expenditures related to purchasing crypto assets should be expensed as they are incurred [35].

According to the *Malaysian Financial Reporting Standards (MFRS) 107*, cash equivalents are short-term, highly liquid investments that are easily convertible. Investments in cryptocurrencies are not equivalent to cash because their prices frequently fluctuate significantly. However, crypto assets are neither financial assets nor equity instruments of another business, such as ordinary shares. A prepaid expense is similar to a crypto asset that grants the holder access to future goods or services rather than the right to cash or another financial asset. However, the classification as a prepayment may not be appropriate if the holder does not intend to take delivery of the underlying goods or services [36].

According to the *IFRIC*, IAS 2 is applicable to cryptocurrencies that are kept for sale in the regular course of business. A company applies IAS 38 to its cryptocurrency assets if IAS 2 is not applicable. IAS 2 on Inventories shall apply if held for sale in the regular course of business and measured at fair value less cost to sell. When retained for capital gains, IAS 38 is applicable and should apply the *Cost or Revaluation Model* because of its infinite lifespan; as such, amortization is not necessary [37].

Factors to consider when determining accounting treatment

Cryptocurrencies and NFTs are included in the definition of property in *Canada's Income Tax Act*. The type of income generated by a property upon sale determines whether it is classified as a capital asset or inventory. As such, one must first determine the nature of the income before defining the property. Factors to be considered in determining the accounting treatments include the following:

Transaction Frequency: Refers to the history of significant buying and selling of NFTs or a rapid turnover of NFT.

Duration of Ownership: Holding NFTs for very short periods of time indicates business transactions rather than capital investment.

Knowledge of NFT Markets: Familiarity with NFT markets, such as a business description, is more likely with more experience or knowledge of NFT markets.

Connection to the Taxpayer's Other Work: If a taxpayer's employment or other business includes NFT transactions (or similar dealings), this indicates a business [38].

Whether an individual is a buyer, seller, creator or marketplace, the type of funds used to purchase the NFT can have a significant impact on how the NFTs are treated in accounting. NFT creators have started adding more rights and obligations to their NFTs. The accounting implications of these additional rights and duties, including the timing of revenue recognition and the distribution of the transaction price, must therefore be considered by creators. Depending on the type of transaction and the parties involved, the creation and sale of NFTs must be evaluated using the applicable accounting rules. Enterprises must therefore make sure that the technical accounting professionals selected are familiar with NFTs, digital assets and sophisticated technical accounting. [39]

Regulatory framework for digital assets

In recent years, one of the most discussed subjects has been the necessity for cryptocurrency legal frameworks, and a number of countries are still working on developing one. From a regulatory perspective, the road to digital assets is not easy. Many initiatives thrive around the world, resulting in different regional perspectives and interpretations. The European Union has taken note of new crypto laws (EU). *Markets in Crypto Assets (MiCA)* is a comprehensive regulatory framework for digital assets and service providers [40]. Digital assets are gaining prominence in the financial services industry. The market capitalization of digital assets such as cryptocurrencies and NFTs was estimated at $1.95 trillion in mid-August 2021 and valued at $2.14 trillion after just one month and is still growing. According to the SEC, digital assets are defined as assets issued and transmitted using distributed ledger or blockchain technology [41].

USA digital asset regulations

Digital assets not only have opportunities to strengthen the financial system globally but also pose risks in the area of crypto market price volatility. President Biden's March 9, 2022, Executive Order (EO) to ensure the accountable growth of digital assets is the first step toward addressing the risks, realizing the potential benefits and their underlying technologies. Government agencies have collaborated to create frameworks and policy suggestions to facilitate the six main points identified in the EO [37]. The key points in the EO include the following:

Protecting Consumers and Investors: Risks associated with digital assets are significant for investors, enterprises and consumers due to high volatility. According to FBI reports financial losses from digital asset fraudulent acts increased by about 600% in 2021 compared to the previous year [37]. *The Treasury Crypto Assets Report* urges agencies to investigate illegal activities and provide supervisory guidelines in order to mitigate risks [42]. In addition, it is advised that investors and consumers in the United States become better educated in order to acquire reliable information on crypto assets to set up mitigation techniques designed to curtail money laundering and terrorism financing [42].

Promoting Safe and Affordable Financial Services: All Americans should benefit from the digital economy and to achieve this, financial services must be created that are trustworthy, dependable, inexpensive and available to everyone. Certain digital assets could speed up payments, increase financial service accessibility and benefit underserved customers. Furthermore, by information exchange and the promotion of a variety of data sets and analytical tools, the Treasury will collaborate with international organizations like *The Financial Stability Board (FSB)* and *The Organization for Economic Co-operation and Development (OECD)* to improve their ability to recognize and address cyber hazards [37].

Combat Illegal Funding: The United States has taken the lead in implementing its anti-money laundering and combating the financing of terrorism (AML/CFT) framework in the digital assets sphere. It is in the best interests of the country to reduce financial crimes by legislation, supervision, legal action and the use of other US government powers [37].

Leadership and Competitiveness in Global Finance: Global standard-setting organizations are now developing regulations, guidelines and policies for digital assets. In order to ensure that countries' financial, legal and technological infrastructures respect fundamental values like data privacy, financial stability and human rights, the United States collaborates with nations that are still developing their digital asset ecosystems.

Financial Inclusion and Responsible Innovation: Over half of the top 100 financial technology companies in the world as of 2022 are based in the United States, many of which provide services related to digital assets. The US government has traditionally been instrumental in fostering ethical private-sector innovation. It supports cutting-edge research, supports businesses' worldwide competition, helps them comply with regulations and collaborates with them to reduce the negative effects of technology innovation [37].

The US Treasury Department in response to the Executive Order and two Senate hearings examined key elements of web3's (also known as Web 3.0, an idea for a new World Wide Web that incorporates token-based economics, blockchain technologies and decentralization) policies and regulations. Emphasis on consumer protection and enforcement, gaps in the current policy landscape that need to be filled, the continued need for exploration of a *US Central Bank Digital Currency (CBDC)*, the collaboration essential to realize the benefits of digital assets and non-financial applications of web3 remain under the radar [42].

European regulatory developments

A draft framework for digital operational resilience and proposed legislation on crypto assets are included in the *Digital Finance Package* that the European Commission released on September 24, 2020. Regulators have attempted to reference and formulate guidelines for the digital space, despite concerns that increased regulation in this area could suffocate digital financial innovation along with the benefits it brings. The intention of the regulation is to create controls, market stability and investor protection regarding digital assets and digital data storage. It includes a suggested pilot program for market infrastructures using distributed ledger technology (DLT). The package aims to ensure risk minimization while fostering innovation and competitiveness in digital finance.

Germany established a specific crypto-asset regulatory framework in 2020. This presented crypto resources as another class inside the meaning of monetary instruments carrying them into the administrative extension. Additionally, it led to the introduction of a licensing requirement for services like custody, administration and security for crypto assets. The regulated institutional *Spezialfonds* (a German open-ended domestic special Alternative Investment Fund with fixed investment terms) will permit cryptocurrency as an asset class on August 1, 2021 [43].

MiCA regulatory framework for digital assets

The European Commission's Proposal for Regulation of the Market in Crypto-Assets (MiCA) is a regulatory framework developed in 2018. The objective of MiCA is to provide legal assurance for crypto assets that are not covered by existing EU financial services legislation, to establish uniform rules for crypto-asset service providers and issuers at the EU level, and replace existing national frameworks related to crypto assets [43]. Once adopted, MiCA will be applicable to all member states of the European Union (EU). It offers licensed services within the EU and suggests a legal framework for assets, markets and service providers that are not currently governed on an EU level. [44]. Any crypto asset that falls under the scope of MiCA shall be subject to MiCA's regulation of standard services. Services that fall under the category of crypto asset include offering advice on crypto assets and taking care of crypto assets on behalf of third parties. In order to offer crypto-asset services, service providers must have authorization and meet specific requirements as set out by MiCA. It is not appropriate to think of the Markets in Crypto-Assets legislation as a stand-alone concept. *The Digital Operational Resilience Act (DORA)* and the *DLT Pilot Regime* are also used as part of the European Commission's digital finance agenda [45].

Current tax considerations in cryptocurrencies and NFTs

The advent of blockchain comes at a time when many are debating whether the present tax code, which was created for brick-and-mortar trade, is still appropriate in this virtual and technological age. It has become clear that business models and technology are developing at a breathtaking rate, while politicians seem far behind on policies [46]. According to a 2015 *World Economic Forum* (WEF) survey, 73.1% of participants anticipated that by 2023, tax authorities throughout the world would have started using blockchain to collect taxes [47]. However, operational inefficiencies and loopholes in the current global taxation systems have resulted in the loss of billions of dollars each year when handling complicated transactions [48]. As such, nation states typically experience these losses instead of corporations [47]. In order to combat tax fraud by tracking down and matching data and enhancing reporting, some countries' tax authorities are now researching the use of blockchain technology. Most consulting firms have already started research into how blockchains may strengthen the tax system and possibly eradicate VAT fraud [48]. The cost of complying with tax laws could be significantly reduced for both corporations and governments. The most well-known application of blockchain technology is found in cryptocurrency transactions. However, it's not entirely clear how to classify such transactions. Should transactions be viewed as the performance of services, the transfer of goods, or something else entirely? [46]. A few instances of cryptocurrency transactions that can be subject to taxation include trading one cryptocurrency for another, mining cryptocurrency and using cryptocurrency to pay for goods and services [49].

Uncertainty in foreign transactions and rising compliance costs for businesses are results of the global push for tax transparency and anti-tax avoidance policies. Blockchain has the capability to streamline and automate tax compliance and transparency. If companies have blockchains in place that record all expenses and every transaction, tax authorities can access this data in real time and calculate and enforce taxes in real time, reducing fraud and avoidance opportunities [46]. The extraction of real-time information would be made easier, and data security and management would be improved by storing data on a blockchain [49]. Real-time, immutable, decentralized, trusted and transparent transactions are qualities of blockchain technology important for the global tax system. A Bitcoin transaction may be categorized as a security, currency or reward token depending on the jurisdiction, with each classification having its own set of tax implications. If categorized as a reward scheme, a different set of tax regulations will apply than if categorized as a commodity [46]. Additionally, it's probable that some cryptocurrency-specific charges may be accepted as a deduction (such as transaction fees, advertising cost, contract drafting cost). Tax fraud could be decreased by using blockchain to trace where and when taxes have been paid. The effective aggregation of data required for tax authorities across numerous countries can also benefit taxpayers [49].

Regulatory frameworks in various countries

In many areas, including tax policy, applications based on distributed ledger technologies like blockchain present challenges for policymakers. Tax administrations and policymakers face difficulties because of the increasing use, trade and market capitalization of these assets and the rapid technological evolution of their features. Guidance on how to deal with crypto assets is being issued by some nations in response to these difficulties. However, no globally accepted guidelines have yet been published, as various issues still need to be investigated, especially on hybrid tokens (a token that consists of two or more crypto derivatives), or tokens that combine aspects of both utility and payment security tokens. Several nations have yet to provide advice on how to categorize cryptocurrency assets under their own laws [20].

Some countries regulations on cryptocurrency assets include the following:

USA: The Internal Revenue Service of the United States has classified cryptocurrency as property. As such, transactions involving cryptocurrency are subject to the standard tax laws that apply to property transactions. Capital gains or losses must be recognized when cryptocurrency is sold [49].

India: Prior to February 2022, India's income-tax legislation did not contain any provisions addressing the taxation of digital assets like cryptocurrencies or NFTs. Subsequently, policies pertaining to the taxation of digital assets were announced by the Indian finance minister on February 1, 2022, and went into effect on April 1, 2022. The new class of digital assets was created by the *India Finance Act 2022* to give clarity on whether to classify the income as a capital gain, business income, or income from other sources. Also, the effect of losses on tax liability is unpredictable due to the volatility that characterizes these developing technologies. Acquisition cost is deductible against gains resulting from the sale of NFTs and taxed at a rate of 30%. As such, when an entity sells other NFTs and incurs losses, the losses cannot be offset against gains. Furthermore, exchanges that facilitate NFT transactions must withhold taxes at a rate of 1% from payments made to resident sellers under certain financial criteria [50].

Ireland: The Annual Report 2021 and *Annual Performance Statement 2022 of the Central Bank of Ireland*, released on May 30, 2022, provide a useful summary of the Central Bank's priorities for the coming year as well as the rate of technological innovation. The Irish Central Bank will continue to develop its regulatory framework to reflect advancements in distributed ledger technology and crypto assets [40].

Taxation approaches

Some key tax issues that crypto assets have brought to governments' attention for direct and indirect tax purposes are how should crypto-asset income be treated. If crypto assets are regarded as property, should they be subject to capital gains taxes or net wealth taxes? Although virtual currencies are typically regarded as intangible assets, IFRIC notes that when held for sale in the ordinary course of business, these assets should be accounted for as inventory in accordance with IFRS IAS 2. According to *Smith* and *Castonguay*, several major accounting firms propose to classify virtual currencies as intangible assets other than goodwill, rather than establishing a new asset class. Most tax administrations have adopted this strategy as they have not yet established specialized and confined tax regimes to tax the creation, mining, exchange and storage of virtual currencies [20].

Income and Value Added Tax (VAT) treatments for virtual currencies

For the purposes of the Income Tax Act, the *CRA* classifies cryptocurrencies as commodities. Depending on the circumstances, any income from cryptocurrency transactions is classified as either commercial income or capital [21]. The *Internal Revenue Service of the United States* has classified cryptocurrency as property, and transactions involving cryptocurrency are subject to the standard tax laws that apply to property transactions. As such, capital gains or losses must be recognized when cryptocurrency is sold [49]. Losses are also considered as business losses or capital losses depending on whether they qualify as capital gains or business revenue. Taxpayers must determine whether cryptocurrency-related activities generate income or capital since doing so has an impact on how the revenue is handled for income-tax reasons. The purchase and sale of cryptocurrencies are not always done by taxpayers who are engaged in business [21].

The CRA generally holds that the fair market value is the price, stated in dollars, that a willing buyer and a willing seller, both of whom are educated, informed and acting independently of one another, would agree to in an open market. A digital wallet that contains multiple cryptocurrencies is deemed to have multiple digital assets, each of which must be appraised independently. As of this writing, there are no taxes associated with holding or owning cryptocurrencies. Selling, giving, trading and exchanging cryptocurrencies including exchanging one cryptocurrency for another are all forms of cryptocurrency transactions.

The proceeds from selling cryptocurrencies might be regarded as company income or a capital gain. Some instances of cryptocurrency businesses are cryptocurrency mining, exchanges for cryptocurrencies, including ATMs and cryptocurrency trading [21]. As such, the question is how should VAT systems deal with these assets' creation, acquisition, holding and transfer, and what policy implications do the various tax treatment options have? [20].

EU and other jurisdictions

The growing usage of cryptocurrencies in day-to-day transactions has piqued the interest of many governments and central agencies, prompting them to intervene and levy taxes [51]. Almost all governments lack explicit tax rules governing the taxation of cryptocurrencies. The tax treatment is based on basic principles and any Tax Authority advice. Some jurisdictions' Tax Authorities have released more detailed guidelines than others, while some jurisdictions have yet to publish any information. A few of these jurisdictions include:

Belgium: When a private individual investor makes a speculative investment, realized gains are taxed at 33% plus local levies. Gains on such investments may be excluded from tax if they are not speculative and do not include any professional activity, while losses are not tax deductible. Profits from currency exchange movements (including cryptocurrencies) are included in taxable profits and losses are deductible for enterprises subject to the usual corporate tax regime. The *Belgian Minister of Finance* certified that the trade of cryptocurrency is VAT-free. Portfolios of cryptocurrency are not taxable assets under the new Belgian portfolio tax [52].

France: One-time profits made on cryptocurrencies are treated as capital gains realized on the sale of intangible assets and are taxed at a flat rate of 19% plus 17.2% social contributions (a total rate of 36.2%). Profits from cryptocurrency mining are handled as commercial and business income in line with the tax schedule. Profits from cryptocurrencies including currency speculation and currency mining are taxable to corporations under the general corporation tax regime for profits and losses. As a delivery of services, sales from cryptocurrency mining are subject to VAT. Purchases of goods or services made using cryptocurrency are also subject to French VAT. The sale of cryptocurrencies is exempt from VAT if done for the purpose of generating ongoing revenue [52].

United Kingdom: Individuals who hold cryptocurrencies as an investment will be taxed on capital gains. Individuals who trade cryptocurrencies will be taxed on their gains as income. Profits or losses from currency exchange movements, including cryptocurrencies, are taxed as income for corporations subject to corporation tax. The revenue generated by crypto mining operations is normally exempt from VAT. Miners' revenue from other activities, such as the supply of services in conjunction with the verification of specified transactions for which specific charges are paid, is also VAT-exempt [52].

Many governments have imposed virtual currency reporting requirements on institutions and corporations to combat money laundering and other criminal financial behavior. Canada and the United States are two nations that have recently established virtual currency disclosure requirements on cryptocurrency exchanges by court application. [53].

Impact of taxation on cryptocurrencies and NFTs

According to Dennis Post, an Ernest Young Blockchain Tax expert, the tax implications of acquisition, ownership and sale vary greatly between jurisdictions, thereby creating ambiguity, complexity and risk that individuals and corporate tax teams must negotiate. The fact that digital assets skip intermediaries and do not have the same tax reporting requirements as traditional investments is a significant challenge for governments globally [54].

Several nations are yet to provide advice on how to categorize cryptocurrency assets under their own laws. The difficulty of defining the tax treatment and the rapid nature of changes may partly be the cause for lack of guidance [20]. Furthermore, blockchain adds a fresh perspective to two long-standing tax questions: where did the transaction occur for tax purposes and who is the taxpayer? Using smart contracts for transactions on the blockchain makes it difficult to pinpoint the location of the last transaction therefore permits anonymity. What is the taxable cost of this sort of transaction? It becomes more complicated if transactions are paid in digital money like Bitcoin. Although this is the early phase of the development of the blockchain, it is essential to begin defining these questions now to shape the policies that can produce the solutions [46]. Concerns over consumer protection, privacy and hazards to the economy, law and ecology have been sparked by the emergence of NFTs. Consumers in the NFT ecosystem face several hazards, and certain NFT markets and digital wallets are lacking in the fundamental security measures needed to shield them from fraudulent actions [55].

Emerging issues

The state of *Washington* made an announcement in July that it was the first state to tax NFTs in the United States. Enforcement of the regulations may be even more challenging, given its complexity. The most likely challenge for tax authorities is to determine whether digital, decentralized and location-less goods like NFTs were sold in Washington. In the past, buyers and sellers of NFTs were not avoiding taxes but rather taking advantage of the absence of regulations. Given the fluctuating value of cryptocurrencies, the issue of determining the taxable value of a person's NFT is addressed in the Washington guidelines. According to the regulations, if a seller receives cryptocurrency in exchange for an NFT, the value of the cryptocurrency must be converted into US dollars at the time of sale [56]. Despite Washington's current sales tax guidance, the decentralized nature of the crypto space may make it difficult to enforce the rules. The self-custody wallet-based marketplaces, such as *OpeanSea* and *SupreRare*, do not have access to information that can be used to locate buyers and sellers. Public perception could be negatively impacted by actively collecting this data for sales tax compliance purposes. It will be nearly impossible to enforce the rules at the marketplace level if the marketplace is truly decentralized and operated as a DAO by a community of wallet holders operating as a pseudo-anonymous community [57]. It might enable the tax authorities to obtain transaction-level data in a timely manner to discover tax leaks. Blockchain might also be a useful tool to revolutionize transfer pricing. Additionally, blockchains could increase the accuracy of accounting and transfer pricing information, which could lower compliance expenses [49].

Tax proposals around the world

Over the past few years, cryptocurrency has seen a steady rise in popularity. People all over the world are increasingly using cryptocurrencies as investment opportunities and payment methods. As such, several nations have proposed tax policies for cryptocurrencies with some exempting cryptocurrencies from all taxes and others making them taxable in the same way as stocks and other property.

> *United Kingdom*: The United Kingdom doesn't think of cryptocurrencies as money or currency but rather as property. Cryptocurrencies have been divided into three distinct categories by *Her Majesty's Revenue and Customs (HMRC)*, the UK tax authority, as utility tokens, security tokens and exchange tokens. These tokens are taxed in the

same way now, but the way they are treated is likely to change in the future. Taxes are levied as capital gains on any profits made from cryptocurrency trading and sales. In the United Kingdom, capital gains are pooled; simply put, a pool is the average cost of all the coins in it. Therefore, the cost of the pool must be used to calculate the capital gain or loss whenever cryptocurrency is sold or disposed. Mining and stake profits are subject to ordinary income tax [58].

Portugal: On August 27, 2019, *Cointelegraph en Espaol* reported that Portugal's tax authority clarified that cryptocurrency trading and cryptocurrency payment transactions will not be taxed in the country. According to the document, crypto users are exempt from income tax and the exchange of crypto for fiat currency is VAT-free [59]. Portugal is currently the only nation in Europe to have taken such a position. The statement is in line with Portugal's closed tax system, which allows only explicitly listed items like stocks, bonds and so on to be taxed. As such, crypto traders find Portugal to be a lucrative country.

Germany: Since 2013, Germany has officially recognized Bitcoin as private money. A capital gains tax – currently 25% – is levied on Bitcoin owners. However, only cases in which Bitcoin profits are realized within a year of purchase are subject to this tax. If the holders of the cryptocurrency keep it for more than a year, they are not subject to capital gains tax [58].

Switzerland: Qualified individuals in Switzerland who buy, sell or hold cryptocurrencies for personal gain are exempt from capital gains tax. However, mining income, which is considered self-employment income, is subject to income tax. Profitable crypto trading by qualified professionals is subject to corporate tax, and wages paid in Bitcoin must be reported for income-tax purposes.

Japan: In Japan, Bitcoin is largely acknowledged as a form of payment. As of July 1, 2017, the *Payment Services Act* declared that the sale of Bitcoin will be exempt from the Consumption Tax. Virtual currencies, such as Bitcoin, are recognized in Japan as assets that may be transferred digitally and used for payment. Generating income with Bitcoin is considered as business revenue; as such, owners must pay income and capital gains taxes [58].

Future prospects for cryptocurrencies and NFTs

According to a report by *Allied Market Research* (AMR), a market research consulting and advisory firm, the worldwide cryptocurrency industry, which was valued at $1.49 billion in 2020, is projected to grow to $4.94 billion by 2030. The cryptocurrency industry is forecasted on a compound annual growth rate (CAGR) of 12.8% between 2021 and 2030 [60]. Blockchain technology and crypto-asset service providers may increase the effectiveness of remittances and overseas payments for both commercial and private clients. When used in conjunction with the smart contract and foreign exchange features, crypto asset and DLT applications may be able to eliminate the need for correspondent banks and additional transaction fees by providing participants with a single source of reference data to help them synchronize the movement of physical goods, information and financing. Platforms, protocols and applications for crypto-assets launched services that would enable users to designate payment streams for workers, suppliers, contractors and other beneficiaries [61].

Global impact of the cryptocurrency market on the economy

An inflow of attention will continue to drive the industry's growth for some time. Some experts believe that larger, global corporations will accelerate adoption even further in the second half of this year. Further institutional adoption could result in more use cases for everyday users, influencing crypto prices. Experts are divided on the subject, with some calling it a bubble and others claiming that the technology underpinning NFTs provides actual value. Meanwhile, producers and artists argue that this is the next form of monetization. Some analysts predict that the NFT market will continue to suffer as a result of the falling price of cryptocurrencies, as well as other macroeconomic variables such as inflation and increasing interest rates [62]. Real-time settlement is a goal of crypto-asset technologies, which could reduce trading risks and costs compared to secondary markets for most traditional non-derivative financial products. In the absence of widespread adoption of real-time settlement, crypto-asset technologies may present an opportunity to shorten the settlement cycle for US securities markets [63].

Organizations and digital assets financial reporting

Since Bitcoin's inception in 2009, crypto assets have developed and diversified significantly, evolving to suit a variety of objectives and economic activities. Although the majority of crypto assets are intended to be used as a means of exchange, as of 2022 less than 30,000 shops worldwide accept crypto assets as payment [63].

Some organizations are testing various forms of crypto assets in order to increase efficiency in areas such as cross-border commerce [64]. In 2021, *Tesla* and *MicroStrategy*, the two most visible corporate users of cryptocurrencies thus far, both saw Bitcoin as an intangible asset with an infinite lifespan. They report it as a digital asset on their balance sheets, and because it has an infinite life, there is no amortization [65]. Tesla reported purchasing $1.5 billion worth of Bitcoin for its balance sheet in its 10-K filed with the Securities and Exchange Commission (SEC). It also stated that it might acquire and hold digital assets occasionally or for a longer period of time [66]. Companies must report Impairment Losses on indefinite-lived intangible assets when their value declines under US GAAP, but they cannot revalue them outside of merger and acquisition (M&A) transactions. In practice, realized gains and losses are always shown on the income statement [65]. Tesla can only earn profits when it sells Bitcoin, but it must include unrealized losses in its net income. If Tesla continues to buy cryptocurrencies as part of its plan, this might account for a sizable portion of the company's cash reserves.

Galaxy Digital is a financial services organization that specializes in digital asset management. It trades, invests and mines cryptocurrencies, as well as provides traditional asset management and investment banking services. Galaxy Digital reports realized and unrealized gains and losses on its Income Statement because it views crypto as investments or financial assets. Digital Assets ought to be regarded as non-core unless the company's primary business is related to crypto. The good news is that crypto is still not widely used, and only a few well-known businesses appear to be using it [66].

The largest flexible space provider, *WeWork* (a coworking space provider that offers both physical and virtual shared workspaces), has declared that it will start supporting a new economy by accepting payment in a few specific cryptocurrencies. By using cryptocurrencies for both inward and outward transactions in collaboration with *BitPay* and *Coinbase*, the business will increase its flexibility. WeWork will take Bitcoin (BTC), Ethereum (ETH), USD Coin (USDC), Paxos (PAX) and several other cryptocurrencies through BitPay, a cryptocurrency payment service provider. The money will be kept on WeWork's balance sheet and when appropriate, the business will pay tenants and other partners in cryptocurrencies via Coinbase, a WeWork member and the biggest US Bitcoin trading platform [67]. The FASB and IASB should collaborate closely to create a harmonized crypto asset accounting model. The entity's purpose and objective for cryptocurrency accounting should be reflected in the presentation and disclosure of cryptocurrency assets and transactions [68].

CONCLUSIONS AND RECOMMENDATIONS

As discussed, cryptocurrencies and NFTs are examples of digital assets providing a more cost-effective operational approach by eliminating intermediary parties' commissions. Cryptocurrencies are very volatile as government policies and regulations – in addition to other market forces can cause instability in the system. However, its acceptance rate is low because cryptocurrencies are still relatively new. Several countries are yet to put in place regulatory frameworks to adopt cryptocurrencies. The valuation of cryptocurrencies is difficult due to a lack of comparable trades, price differences between buy and sell orders and the exchanges used for the trade.

NFTs are tokens that establish ownership of an underlying asset in a cryptographically secure manner. However, the value of NFTs is solely dependent on their visual and sentimental appeal. The fundamental distinction is that the value of cryptocurrency is exclusively determined by its economic function as a currency or investment. NFTs, on the other hand, have economic and non-economic value.

Any monetary or non-financial item that cannot be physically handled is considered an intangible asset. A separable asset or one that is derived from contractual or other legal rights makes it an identifiable non-monetary asset. An organization typically measures an intangible asset at cost less accumulated amortization after initial recognition. In rare instances where fair value may be established by reference to an active market, the entity may decide to measure the asset at fair value. When it comes to applying standard methods of valuation to crypto assets, there is a healthy amount of skepticism. This has increased due to the tremendous volatility of Bitcoin and other cryptocurrencies. It is still appropriate and applicable to determine these assets' intrinsic value using conventional valuation procedures and principles. As market enthusiasm drives pricing, such strategies should consider the intrinsic volatility or riskiness of these assets. Financial reporting, taxation and investment concerns are the three main factors driving the demand for a thorough approach to crypto-asset valuation.

One of the first considerations to examine when valuing digital assets is the basis of valuation, that is, if the value is for transaction purposes, taxation, financial reporting or some other reason. The valuation foundation and classification in the appropriate framework both have an influence on the eventual method of establishing value. The three widely used valuation methods in line with International Accounting Standards are the market approach, cost approach and income approach. Additional valuation methods that can be applicable to digital assets are asset approach and Mark-to-Market methodology. The Mark-to-Market method involves adjusting an asset's value to reflect the current market value. The method can either be used to measure the fair value of funds that are subject to periodic fluctuations, or it can be used in financial markets to represent the current and fair market value of investments.

Blockchain in accounting has the potential to increase the quality of information that reaches investors by making the accounting data more reliable and timely. In addition, this information would be instantly made available to regulators as well as insiders within the company. Currently, there are no set guidelines on how to account for cryptocurrencies. *CPAs* and accounting firms have requested that the FASB address this rising concern and to consider publishing updated guidelines in line with these new assets. According to the FASB, businesses should track crypto assets using fair value accounting, recording gains and losses in current period comprehensive income, but this decision is not yet final.

In recent years, one of the most discussed subjects has been the necessity for cryptocurrency legal frameworks, and a number of countries are still working on developing one. Several nations have proposed tax policies for cryptocurrencies with some exempting cryptocurrencies from all taxes, while others making them taxable in the same way as stocks and other property. Regulators have attempted to reference and formulate guidelines for the digital space, despite concerns that increased regulation in this area could suffocate digital financial innovation and the benefits it brings. If companies have blockchains in place that record all expenses and every transaction, tax authorities can access this data in real time and calculate and enforce taxes in real time, reducing fraud and avoidance opportunities. A tremendous inflow of attention will continue to drive the industry's growth for some time. Further institutional adoption could result in more use cases for everyday users, influencing crypto prices.

Digital assets over time have continuously experienced growth as evident in the global market capitalization. The valuation and financial reporting of digital assets require an enhanced approach that will be acceptable worldwide by regulators, standard setters and governments of various nations, with respect to the framework and guidelines in the financial statements.

As of now, there are no formal guidelines on how to account for cryptocurrencies [33]. The FASB should provide further advice to ensure uniformity in the treatment and valuation of digital assets. FASB should carry on with a larger digital assets initiative that might address issues with NFTs, stablecoins, CBDCs and any further issues that might surface before the project is complete. Another suggestion is that the FASB consider addressing issues such as emission allowances and renewable energy credits that are maintained for investment or actively traded digital assets [33].

The use of blockchain technology and smart contracts can revolutionize the areas of accounting that deal with transactional assurance and transferring property rights. More attention can be paid to accounting for transactions, thereby decreasing the requirements for reconciliation. By using blockchain technology to optimize many present accounting department operations, the accounting function will become more effective and valuable. However, more focus should be on how transactions are recorded and recognized in the financial statements and how subjective elements like valuations are decided [69].

Organizations may be concerned about potential risks from adopting blockchain in the accounting treatment of financial statements. In order to investigate further applications of blockchain-enabled accounting, more study is required. Further research might attempt to address this – expanding the field of blockchain on financial statements – therefore ensuring integration into the financial system. Blockchain needs to be developed, standardized and enhanced in order to overcome technical, organizational and regulatory barriers [70].

REFERENCES

[1] Vincent, M. (2021, October 21). *What are digital assets and how does blockchain work?* Financial Times. Retrieved October 12, 2022, from https://www.ft.com/content/2691366f-d381-40cd-a769-6559779151c2

[2] Härdle, W. K., Harvey, C. R., & Reule, R. C. (2020). Understanding cryptocurrencies. *Journal of Financial Econometrics*, *18*(2), 181–208. https://doi.org/10.1093/jjfinec/nbz033

[3] Official, A. M. C. (2020, August 28). Key characteristics of cryptocurrency and why do they matter to you. *Medium*. Retrieved November 12, 2022, from https://medium.com/the-capital/key-characteristics-of-cryptocurrency-and-why-do-they-matter-to-you-5f33e483a40f

[4] Bunjaku, F., Gjorgieva-Trajkovska, O., & MitevaKacarski, E. (2017, December 5). *Cryptocurrencies – Advantages and Disadvantages*. ISSN.

[5] Nibley, B. (2022, October 28). *14 benefits of cryptocurrency in 2022*. SoFi. Retrieved September 30, 2022, from https://www.sofi.com/learn/content/benefits-of-crypto/.

[6] Rice, M. (2019). *Cryptocurrency: History, Advantages, Disadvantages, and the Future. Running Head: Cryptocurrency: What is it?* Retrieved November 14, 2022, from https://core.ac.uk/download/pdf/288850271.pdf

[7] The Law Library of Congress. (2018, June). *Regulation of Cryptocurrency Around the World*. The Law Library of Congress.

[8] Caginalp, C., & Caginalp, G. (2018). Valuation, liquidity price, and stability of Cryptocurrencies. Proceedings of the National Academy of Sciences, 115(6), 1131–1134. https://doi.org/10.1073/pnas.1722031115

[9] EQVISTA. (2022, September 20). *NFT valuation - everything you need to know*. Eqvista. Retrieved November 8, 2022, from https://eqvista.com/company-valuation/valuation-crypto-assets/nft-valuation/

[10] Wang, Q., Li, R., Wang, Q., & Chen, S. (2021). Non-Fungible Token (NFT): Overview, Evaluation, Opportunities and Challenges (Tech ReportV2). *ArXiv*. https://doi.org/arXiv:2105.07447v3

[11] Geroni, D. (2022, August 15). *Understanding the attributes of non-fungible tokens (NFTs)*. 101 Blockchains. Retrieved November 13, 2022, from https://101blockchains.com/nft-attributes/

[12] Geroni, D. (2022, August 15). *The advantages of non-fungible tokens (NFTs)*. 101 Blockchains. Retrieved November 12, 2022, from https://101blockchains.com/advantages-of-nfts/

[13] Edwards, S. (2021, September 18). *NFTs Pros & Cons: The good, the bad, and the ugly*. The Dales Report. Retrieved November 12, 2022, from https://thedalesreport.com/crypto-nfts/nfts-pros-cons-the-good-the-bad-and-the-ugly/

[14] Lidén, E. (2022, June 7). *Potential Advantages and Disadvantages of NFT-Applied Digital Art*. diva-portal. Retrieved November 14, 2022, from https://www.diva-portal.org/smash/

[15] Rehman, W., Zainab, H. e, Imran, J., & Bawany, N. Z. (2021). NFTS: Applications and challenges. *2021 22nd International Arab Conference on Information Technology (ACIT)*. https://doi.org/10.1109/acit53391.2021.9677260

[16] 101 Blockchains. (2022, August 15). *Non-fungible tokens - risks and challenges*. 101 Blockchains. Retrieved November 14, 2022, from https://101blockchains.com/nft-risks-and-challenges/

[17] Lisa, A. (2022, June 29). *NFT vs. crypto: What is the difference?* Nasdaq. Retrieved October 12, 2022, from https://www.nasdaq.com/articles/nft-vs.-crypto%3A-what-is-the-difference

[18] CoinMarketCap. (2022, November 11). *Cryptocurrency prices, charts and market capitalizations*. CoinMarketCap. Retrieved November 12, 2022, from https://coinmarketcap.com/

[19] IFRS. (2022). *IAS 38 Intangible Assets*. Retrieved October 27, 2022, from https://www.ifrs.org/issued-standards/list-of-standards/ias-38-intangible-assets/

[20] OECD. (2020, October). *Taxing virtual currencies: An overview of tax treatments and emerging tax policy issues*. Retrieved October 7, 2022, from https://www.oecd.org/tax/tax-policy/taxing-virtual-currencies-an-overview-of-tax-treatments-and-emerging-tax-policy-issues.html

[21] CRA. (2021, June 26). *Government of Canada*. Canada.ca. Retrieved November 6, 2022, from https://www.canada.ca/en/revenue-agency/programs/about-canada-revenue-agency-cra/compliance/digital-currency/cryptocurrency-guide.html

[22] EY (2019, March 7). *The valuation of crypto-assets*. www.ey.com. Retrieved March 2, 2023, from https://www.ey.com/en_qa/financial-services--emeia-insights/the-valuation-of-crypto-assets

[23] Saidi, A. A. (2019, July 10). *Valuation of crypto assets. A conceptual framework and Case application to the iota token*. GRIN. Retrieved March 2, 2023, from https://www.grin.com/document/494016

[24] Larsen, D., & Weber, L. (2019, November 14). *Digital assets – what are they and can they be reliably valued? - fin tech – united states*. Digital Assets – What Are They And Can They Be Reliably Valued? – Fin Tech – United States. Retrieved March 4, 2023, from https://www.mondaq.com/unitedstates/fin-tech/864262/digital-assets-what-are-they-and-can-they-be-reliably-valued

[25] Trevisi, C., Viscont, R. M., & Cesaretti, A. (2022, June 21). Non-fungible tokens (NFT): Business models, Legal Aspects, and market valuation. MediaLaws. Retrieved November 8, 2022, from https://www.medialaws.eu/rivista/non-fungible-tokens-nft-business-models-legal-aspects-and-market-valuation/

[26] Kaal, W. (2022, February 13). *Digital Asset Valuation*. wulfkaal.medium.com. Retrieved March 4, 2023, from https://wulfkaal.medium.com/digital-asset-valuation-290945e70861

[27] Eqvista. (2022, April 4). *Business valuation: The Market Value Approach*. eqvista.com. Retrieved March 4, 2023, from https://eqvista.com/company-valuation/business-valuation-market-value-approach/

[28] Arad, C., & Nandi, S. (2018, June 15). *Valuation of Cryptocurrencies: A brief analysis of three valuation approaches and their implication*. micobo.medium.com. Retrieved March 3, 2023, from

https://micobo.medium.com/valuation-of-cryptocurrencies-a-brief-analysis-of-three-valuation-approaches-and-their-implication-1525f7fd7a87

[29] CFI Team. (2022, October 4). *Mark to market.* Corporate Finance Institute. Retrieved November 9, 2022, from https://corporatefinanceinstitute.com/resources/valuation/mark-to-market/

[30] Kanter, T. D., Kolovos, J., & Tobin, A. F. (2018, February 28). *Cryptocurrency: Tax guidance, reporting, and more: Crowe LLP.* Cryptocurrency: Tax Guidance, Reporting, and More | Crowe LLP. Retrieved November 8, 2022, from https://www.crowe.com/insights/crowe-financial-services-tax-insights/cryptocurrency-existing-tax-guidance-the-applicability-of-marktoma

[31] Byström, H. (2019, April). *Blockchains real-time accounting and the future of Credit Risk Modeling.* https://www.researchgate.net/publication/332772636_Blockchains_Real-time_Accounting_and_the_Future_of_Credit_Risk_Modeling. Retrieved September 28, 2022.

[32] Alkan, B. Ş. (2021, August 16). *Real-time blockchain accounting system as a new paradigm.* dergipark.org.tr. Retrieved March 3, 2023, from https://doi.org/10.25095/mufad.950162

[33] Jacob, A. (2021, July 3). *A quick guide to accounting for cryptocurrency.* TaxBit. Retrieved November 10, 2022, from https://taxbit.com/blog/a-quick-guide-to-accounting-for-cryptocurrency

[34] Tapscott, A. (2022, November 4). *New Accounting Rules for Crypto Will Change Accounting and the Accounting Profession.* Fortune. Retrieved November 9, 2022, from https://fortune.com/crypto/2022/11/04/crypto-is-about-to-change-bookkeeping-rules-and-soon-the-accounting-profession/

[35] Muir, S. (2022, October 13). *FASB decides how to measure crypto assets.* KPMG LLP. Retrieved November 12, 2022, from https://frv.kpmg.us/reference-library/2022/fasb-takes-on-accounting-for-digital-assets.html

[36] Crowe Malaysia PLT. (2021, June 1). *Accounting for cryptocurrencies in the financial statements.* www.crowe.com Retrieved March 16, 2023, from https://www.crowe.com/my/insights/how-should-cryptocurrencies--be-accounted--for-in-the--financial-statements

[37] The United States Government. (2022, September 16). *Fact sheet: White House releases first-ever comprehensive framework for Responsible Development of Digital assets.* The White House. Retrieved October 12, 2022, from https://www.whitehouse.gov/briefing-room/statements-releases/2022/09/16/fact-sheet-white-house-releases-first-ever-comprehensive-framework-for-responsible-development-of-digital-assets/

[38] TaxPage. (2021, October 25). *Tax implications of buying & selling blockchain NFTs.* TaxPage.com. Retrieved November 12, 2022, from https://taxpage.com/articles-and-tips/canadian-income-tax-implications-of-buying-selling-blockchain-nfts/

[39] Centri Business Consulting, LLC. (2022, January 20). *Accounting Considerations for NFTs.* centriconsulting.com. Retrieved November 14, 2022, from https://centriconsulting.com/news/accounting-for-nfts/

[40] BNP Paribas. (2022, July 5). *Digital Assets: Mapping Regulatory Developments and news across the world.* Securities Services. Retrieved November 6, 2022, from https://securities.cib.bnpparibas/the-road-to-digital-assets/

[41] Phillips, T. (2021, October). *The SEC's regulatory role in the Digital Asset Markets.* americanprogress.org. Retrieved October 13, 2022, from https://americanprogress.org/wp-content/uploads/2021/10/SECs-Regulatory-Role-in-the-Digital-Asset-Markets-1.pdf

[42] Wilkinson, R. (2022, September 27). *What we can learn about the future of Digital Assets Regulation from recent US Government reports.* World Economic Forum. Retrieved October 12, 2022, from https://www.weforum.org/agenda/2022/09/5-takeaways-from-the-u-s-federal-government-s-review-of-digital-assets/

[43] Sandeman, M., & Raschen, H. (2021, July 27). *Digital Assets - European Regulatory Developments: Global Perspectives.* HSBC. Retrieved November 6, 2022, from https://www.gbm.hsbc.com/insights/securities-services/digital-assets-european-regulatory-developments#3-regulatory-developments

[44] Vermaak, W. (2022, July 17). *Mica (updated July 2022): A guide to the EU's proposed markets in Crypto-Assets Regulation.* Sygna. Retrieved October 12, 2022, from https://www.sygna.io/blog/what-is-mica-markets-in-crypto-assets-eu-regulation-guide/

[45] Boucheta, H., & Joseph, A. (2022, July 7). *Mica – markets in Crypto-Assets Regulation memo.* Securities Services. Retrieved November 5, 2022, from https://securities.cib.bnpparibas/markets-in-crypto-assets-regulation/

[46] KPMG. (2018, February 28). *How might Blockchain technology revolutionize tax?* Retrieved October 5, 2022, from https://home.kpmg/mt/en/home/insights/2017/12/how-might-blockchain-technology-revolutionise-tax.html

[47] Hoffman, M. R. (2018). Can blockchains and Linked Data Advance Taxation? *Companion of the Web Conference 2018 on the Web Conference 2018 - WWW'18,* 1–4. https://doi.org/10.1145/3184558.3191555

[48] Deloitte. (2022). *Crypto asset management: Managing the tax expectations gap.* Retrieved from Deloitte.com: https://www2.deloitte.com/content/dam/Deloitte/us/Documents/Tax/us-crypto-asset-management-01.pdf

[49] Kreston Global. (2022, March 09). *Kreston Global.* Retrieved from kreston.com: https://www.kreston.com/article/tax-implications-of-blockchain/#:~:text=Blockchain%20is%20likely%20to%20add%20value%20for%20tax,Blockchain%20could%20be%20a%20potential%20Transfer%20Pricing%20Tool.

[50] Aggarwal, N. (2022, August 17). *To tax or not to tax - the curious case of NFTs.* KPMG. Retrieved October 11, 2022, from https://home.kpmg/in/en/blogs/home/posts/2022/08/to-tax-or-not-to-tax-the-curious-case-of-nfts.html

[51] The European Business Review. (2021, March 21). *Do you have to pay taxes on cryptocurrency in Europe?* europeanbusinessreview.com. Retrieved December 4, 2022, from https://www.europeanbusinessreview.com/do-you-have-to-pay-taxes-on-cryptocurrency-in-europe/

[52] Osborne Clarke. (2018, September 17). *Taxation of cryptocurrencies in Europe: An overview.* osborneclarke.com. Retrieved December 4, 2022, from https://www.osborneclarke.com/insights/taxation-of-cryptocurrencies-in-europe

[53] Taxlaw. (2022, January 27). *Cryptocurrency tax: OECD's 2021 cryptocurrency reporting framework.* taxlawcanada.com. Retrieved December 4, 2022, from https://taxlawcanada.com/cryptocurrency-tax-oecds-2021-cryptocurrency-reporting-framework/

[54] Bradley, I. S. (2022, March 23). *How taxes on cryptocurrencies and Digital Assets will soon take shape.* Retrieved October 12, 2022, from https://www.ey.com/en_gl/tax/how-taxes-on-cryptocurrencies-and-digital-assets-will-soon-take-shape

[55] Congressional Research Service. (2022, July 20). *Non-fungible tokens (NFTs) – Congress.* Retrieved October 12, 2022, from https://crsreports.congress.gov/product/pdf/R/R47189

[56] Mattei, S. E.-D. (2022, September 9). *Washington becomes the first state to Tax NFTs.* ARTnews.com. Retrieved November 14, 2022, from https://www.artnews.com/art-news/news/washington-tax-nfts-regulations-1234638897/#!

[57] Chandrasekera, S. (2022, July 14). *NFT purchases are now being subject to sales taxes.* Forbes. Retrieved November 14, 2022, from https://www.forbes.com/sites/shehanchandrasekera/2022/07/13/nft-purchases-are-now-being-subject-to-sales-taxes/?sh=55bd6b0c9cf9

[58] Deribit. (2021, June 11). *Cryptocurrency tax laws around the world.* Deribit Insights. Retrieved November 13, 2022, from https://insights.deribit.com/industry/cryptocurrency-tax-laws-around-the-world/

[59] Partz, H. (2019, August 29). *Portugal Tax Authority: Bitcoin trading and payments are tax-free.* Cointelegraph.com. Retrieved November 13, 2022, from https://cointelegraph.com/news/portugal-tax-authority-bitcoin-trading-and-payments-are-tax-free

[60] PRNewswire. (2021, August 24). *Cryptocurrency market to reach $4.94 billion, globally, by 2030 at 12.8% CAGR: Allied Market Research.* Cision PR Newswire UK provides press release distribution, targeting, monitoring, and marketing services. Retrieved October 11, 2022, from https://www.prnewswire.co.uk/news-releases/cryptocurrency-market-to-reach-4-94-billion-globally-by-2030-at-12-8-cagr-allied-market-research-836503660.html

[61] U.S. Department of the Treasury. (2022, September). *Crypto-Assets: Implications for Consumers, Investors, and Businesses.* home.treasury.gov. Retrieved November 14, 2022, from https://home.treasury.gov/

[62] Gailey, A., & Haar, R. (2022, October 31). *Future of cryptocurrency in 2022 and beyond | nextadvisor with Time.* Time. Retrieved November 13, 2022, from https://time.com/nextadvisor/investing/cryptocurrency/future-of-cryptocurrency/

[63] U.S. Department of the Treasury. (2022, September). *Crypto-Assets: Implications for Consumers, Investors, and Businesses.* home.treasury.gov. Retrieved November 14, 2022, from https://home.treasury.gov

[64] Bains, P., Ismail, A., Melo , F., Sugimoto, N. (2022, September 26). *Regulating the crypto ecosystem: The case of unbacked crypto assets.* imf.org. Retrieved November 13, 2022, from https://www.imf.org/en/Publications/fintech-notes/Issues/2022/09/26/Regulating-the-Crypto-Ecosystem-The-Case-of-Unbacked-Crypto-Assets-523715

[65] Brian DeChesare, Brian. "Cryptocurrency Accounting: Why Net Income and the P / E Multiple Have Become Even More Useless." *Mergers & Inquisitions*, 7 Dec. 2022, https://mergersandinquisitions.com/cryptocurrency-accounting/

[66] Hake, Mark R. "7 Stocks of Companies That Have Large Cryptocurrency Investments." *InvestorPlace*, 12 Apr. 2021, https://investorplace.com/2021/04/7-stocks-of-companies-that-have-large-cryptocurrency-investments

[67] Business Wire. (2021, April 20). *WeWork starts utilizing cryptocurrency as form of payment.* WeWork Starts Utilizing Cryptocurrency as Form of Payment. Retrieved October 11, 2022, from https://www.businesswire.com/news/home/20210420005667/en/WeWork-Starts-Utilizing-Cryptocurrency-as-Form-of-Payment

[68] KPMG. (2022, August 8). *Crypto and other Digital assets: Hot topics.* KPMG LLP. Retrieved November 13, 2022, from https://frv.kpmg.us/all-topics/crypto-currency-digital-assets/crypto-currency-digital-assets-hot-topics.html

[69] *Blockchain and the Future of Accountancy.* ICAEW. (2023). Retrieved March 17, 2023, from https://www.icaew.com/technical/technology/blockchain-and-cryptoassets/blockchain-articles/blockchain-and-the-accounting-perspective

[70] Han, H., Shiwakoti, R. K., Jarvis, R., Mordi, C., & Botchie, D. (2022, November 24). *Accounting and auditing with blockchain technology and Artificial Intelligence: A literature review.* International Journal of Accounting Information Systems. Retrieved March 17, 2023, from https://www.sciencedirect.com/science/article/pii/S1467089522000501

Part II

Blockchain benevolent use cases

Chapter 7

The use of self-sovereign identity in blockchain systems

Adebimpe Onamade and Ebubechi Okpaegbe

BACKGROUND

The global economy has reached a historical tipping point due to the digital revolution and the availability of multiple channels that facilitate information sharing and interpersonal transactions [1]. Gradually, the digital world has transformed into a virtualized version of the real world, simplifying communication [1]. While the internet was initially built for information exchange, it has grown in economic significance, forming the basis for a safe, sustainable, and trustworthy digital identity in our digital society and economy [2]. According to Kim Cameron, Microsoft's Chief Architecture Officer for Identity, "*the Internet was built without an identity layer,*" leaving users unaware of who or what they are connected to, leading to limited usage and increased vulnerability to threats. If not addressed, this could result in a surge of theft and fraud, eroding public trust in the internet. Therefore, it is crucial to resolve online identity challenges to safeguard the future of the internet [3, 4].

Digital identities are the online equivalents of physical identities, encompassing unique characteristics that establish and verify a person's identity [2]. Similar to a physical ID card like a passport or driver's license, a digital identity represents a user's online identification, incorporating their traits or features [5]. Christopher Allen categorizes the evolution of digital identity into four stages [5]. Over time, the concept of digital identity has evolved from a centralized system to a more decentralized approach, encompassing stages such as centralized identity, federated identity, user-centered identity, and self-sovereign identity (SSI) [4, 6].

Centralized Identity Management Model (IDM 1.0)

The centralized IDM model emerged between 1990 and 2000 when online services became the issuers and custodians of digital identities. Certificate authorities began issuing official digital certificates to establish trust in online services. However, the control and ownership of data remained with the online service providers rather than the users [2]. The centralized identity model, also known as account-based identity, represents the original form of digital identity. It applies to various identities and credentials, including government ID numbers, passports, identification cards, driver's licenses, invoices, Facebook logins, and Twitter handles. Establishing an online identity with a website, service, or application involves creating an account using a username and password [3]. In the centralized IDM model, an organization provides credentials to users, usually in the form of a username and password, enabling access to their services. The organization stores and manages the user's personal information, and the trust relationship relies on a shared secret. This model necessitates separate credentials for each organization or system users wish to access [4, 7]. An example of this IDM model is the traditional *Microsoft Active Directory Domain* (ADDS) employed in enterprises before Windows Server 2003.

DOI: 10.1201/9781003462033-9

Federated Identity Management Model (IDM 2.0)

The federated identity model was prevalent between 2000 and 2008, allowing users to use a single login across multiple services for a seamless customer experience. The main difference from the centralized digital identity is the ability to migrate the digital identity from one online service to another [2]. As a response to the challenges posed by centralized identity, the federated identity model was developed. This model involves a three-way relationship between the user, the identity provider (IDP), and a relying party or organization. Through this model, users can utilize a single identity account to log in and share essential identity information with any application, service, or site that utilizes the IDP [3]. *Naik and Jenkins* [7] state that Federated Identity Management (FIDM) provides a solution to the challenges encountered in traditional Centralized Identity Management (IDM) systems. The key benefits of FIDM include organizations being able to delegate identity management to third-party IDPs, who not only secure credentials and identity information but also streamline business operations. Users can enjoy the convenience of using a single identity credential for accessing multiple systems through single sign-on (SSO). The IDP holds and controls the users' personal information. A commonly used example of FIDM is *Microsoft Azure Active Directory* (AzureAD), utilized in enterprise settings for business-to-business (B2B) and business-to-client (B2C) authentication, with IDPs such as Facebook and Google [4].

Self-sovereign identity management model

With the advent of blockchain technology, a third category of digital identity – SSI – has emerged. In recent times, there has been a growing academic and political debate surrounding SSI, which envisions an independent citizenship identity. The discussion on SSI is relatively new and has made a limited scientific contribution to the research in this field. However, it presents opportunities and prerequisites for implementing an SSI system within the public sector identity scheme [8]. In a nutshell, SSI refers to a type of digital identity in which the user has full control and decision-making authority over who can access their information and under what circumstances [5]. In other words, SSI is based on an identity owned by users, empowering them with control over their own data. SSI has become increasingly necessary given the availability, trust, and privacy concerns associated with commonly used physical and digital IDs [1]. It provides a means to authenticate our identity to websites, services, and applications that require secure connections to access or protect sensitive information. SSI relies on cutting-edge technologies and standards such as cryptography, distributed networks, cloud computing, and smartphones, heralding a new era in digital identity. While the specifics of SSI are still evolving, its well-designed and coherent nature has garnered widespread acceptance [3]. In essence, SSI can be seen as a digital identity framework that places the individual or business at the center of control, enabling them to determine how their identity and information are utilized and shared. Although SSI is not a standalone technology but often discussed in the context of blockchain technology, enthusiasts are actively exploring its implementation [8]. The decentralized nature of blockchain technology is leveraged in the digital, peer-to-peer platform of the SSI system, which is built on the foundation of the trust triangle [5]. The emergence of SSI represents a revolutionary shift in digital identity on the internet.

DISCUSSION

The mechanics of SSI

As mentioned before, SSI is an individual identity that is not dependent on or subject to another power or state. SSI does not mean self-asserted identity rather, because an individual's identity information is coming from a trusted source, other parties are willing and able to rely on it. SSI is not just about people's privacy, security, and personal data control; rather, it applies to any organization or anything that requires identity on the internet. SSI not only aids the resolution of internet identity layer concerns but also shifts control from centralized or federated entities to individual users [3]. According to Naik and Jenkins, self-sovereign IDM model improves upon the federated IDM model by allowing direct connections between users and organizations. The use of a digital wallet solves the problem of user management of their personal information related to identity. The digital wallet retains this data, which is possessed and regulated by the user on their device. In its ecosystem, SSI serves three main functions: Issuer, Holder, and Verifier. Holders receive credentials from issuers and keep them, sharing them with verifiers when necessary. Verifiers receive and check credentials presented by holders. To establish a cryptographically secure and digitally verifiable identity that is solely controlled by its owner, the implementation of SSI leverages new standards, including Verifiable Credentials (VC) and Decentralized Identifiers (DIDs). DID is a distinct and enduring identifier that cannot be taken away from its possessor, unlike temporary identifiers such as domain names, IP addresses, or mobile numbers (Table 7.1) [7].

SSI has gained significant attention in digital identity circles as a new paradigm that places the user at the center, granting them greater control over their identity information [8]. Several key aspects of SSI have evolved, highlighting its value advantages such as allowing users to choose the techniques and types of data they provide to organizations. These characteristics demonstrate how an SSI management system could empower users to fully control the ownership and administration of their identity without relying on other parties. Over time, various concerns regarding user experience, data protection, security, privacy, and interoperability of digital identity have been discussed in different contexts [9]. Central repositories of personal data on issuers' servers are vulnerable, leaving users with no control over how

Table 7.1 Summary side-by-side comparison of the three main identity management models [3, 4, 6, 7]

Feature/Aspect	Centralized ID management	Federated ID management	Self-sovereign ID management
Control	Central authority (Organization or service provider)	Shared control among multiple organizations	Individual (user) control
Data Storage	Centralized database	Distributed across multiple trusted entities	Decentralized, often on blockchain
User Experience	Single sign-on for services provided by the same entity	Single sign-on across different organizations or services	User-managed credentials and identities
Privacy	High risk of data breaches and privacy issues	Improved privacy with shared but controlled data access	Enhanced privacy, user decides what to share and with whom
Scalability	Limited to the capacity of the central authority	Scalable across participating organizations	Highly scalable, reliant on underlying blockchain tech
Interoperability	Typically low, within the same ecosystem only	High, due to agreed-upon standards and protocols	High, designed for wide compatibility
Security	Single point of failure	Reduced risk due to distributed data	Increased security with cryptographic methods
Examples	LDAP, Active Directory	SAML, OAuth, OpenID Connect	Sovrin, uPort, SSI frameworks
Cost	Lower initial cost but higher maintenance and breach costs	Medium initial and operational costs	Variable costs depending on implementation

their data is shared and exposing them to threats [1]. Self-sovereign identities alleviate these concerns through the benefits of encryption and decentralization. This model enables users to have complete control over their identity ownership and administration and allows them to present their data to others for verification as needed [9].

To facilitate faster and simpler issuance of credentials, the SSI system provider needs to be aware of credential exchange. Furthermore, privacy is frequently at risk when third-party login services offer financial incentives for data collection and storage. Establishing a highly secure peer-to-peer digital network among the ID issuer, ID verifier, and ID owner is crucial for SSI. Identity owners can choose which attributes to disclose, granting them full control over interactions with ID verifiers and ensuring transparency in data sharing. Another intriguing aspect of SSI is that users only need to use their wallet password instead of relying on multiple passwords [1]. Unlike other forms of digital identity, SSI does not require constant sharing of sensitive identifiable information. Instead, it relies on three main protocols: verifiable credentials, DIDs, and distributed ledger technology (DLT) [5]. Self-sovereign identification (SSI) systems are envisioned as public utilities, with the underlying blockchain serving as an essential "identity internet." While they primarily rely on non-permissioned public blockchains like Ethereum, the underlying *Hyperledger Indy* blockchain in the *Sovrin Network* is a public but permissioned chain [10].

Recently, there has been a debate regarding the potential distinctions between decentralized identity and SSI. However, these two concepts do not have a clear distinction apart from their underlying ideas. The concept of decentralized identity emphasizes users' ability to regain control over their data, which is crucial for maintaining privacy and identity. Decentralized identities are entirely controlled by the individual without the involvement of a certificate authority, central registry, or identity provider to verify their accuracy [9]. The integration of significant advancements in cryptography, distributed databases, and decentralized networks has led to the development of novel standards for decentralized identification, namely VCs and DIDs. In a decentralized identity model, authentication is based on a peer-to-peer relationship between a user and another party, without any party providing, controlling, or owning the relationship, unlike account-based authentication [3]. Decentralized and SSI differ in terms of registration, identifier registration, and aggregation of identity attributes. Decentralized identity relies on government-issued identity documents such as driving licenses and passports and relies on services that validate identity through trusted third-party credentials. Authenticated identity assertions are stored on a distributed ledger after the verification process and can be subsequently verified by third parties [11]. For decentralized identification to work, users need a mobile identity wallet on their smartphone. The software securely stores credentials and interacts with issuers and verifiers. Decentralized identification techniques do not require the storage of sensitive data (such as personally identifiable information) on the ledger. Instead, the ledger serves as the foundation of trust, allowing verifiers to validate that a credential supplied by a user was cryptographically signed by the appropriate issuer. The blockchain's public-private cryptography tracks the origin of a digital token, while in the case of identity, the DLT infrastructure can be used to track the origin of verifiable credentials. Decentralized systems must integrate with existing identity and access management (IAM) systems, provide appropriate key management (including key recovery and revocation), and adhere to privacy-by-design principles [10]. On the other hand, SSI does not rely on pre-existing documents. Entities can create multiple identities without depending on documents, as identity information can be self-declared or acquired later by collecting credentials from various issuers. In some decentralized proposals, a single identity is allocated to each entity and stored in the blockchain for authentication. SSI enables the creation of an unlimited number of identifiers based on the situation and needs, making it well-suited for IoT implementations where devices can verify each other without a central authority. However, a decentralized identity system must fulfill the principles of SSI, such as control, portability, security, and others [11].

Self-sovereign identity model, protocols, and architecture

Within the SSI architecture, the initial layer utilizes a distributed ledger as a registry for distributed digital identity identifiers. This guarantees that the identity owner controls their private key and no third party can access the identifier. The distributed ledger's non-tampering property is suitable for publishing and managing distributed digital identity data and verifying the legitimacy of credentials.

The second layer combines the public key infrastructure (PKI) system with digital wallets and distributed ledgers to facilitate end-to-end user interaction and perform actions such as issuance, updates, certificate applications, and revocations. Digital wallets function as personal repositories, storing user identity information and verifiable claims (VC) off-chain, thereby returning identity control to the users.

The verifiable claim application layer is the third layer, allowing interaction between the issuer, holder, and verifier. In the SSI model, users have primary control over their identities, with much more authority over their data and information than anyone else. The governance layer is the fourth layer, where legal and business protocols are established in a distributed network to establish human trust. The governance model outlines policies, principles, standards, and terminology that define responsibilities and trusted certificate authorities. The SSI system employs decentralized blockchain technology on a digital, peer-to-peer platform relying on the trust triangle. The SSI triangle of trust consists of three entities: the issuer of the digital ID, the owner of the ID, and the verifier of the ID [2].

Blockchain technology provides a distributed trust environment essential for implementing SSI. A blockchain-based decentralized identity management system combines distributed data storage, point-to-point transmission, encryption security, consensus confirmation, and other features to effectively address issues related to identity verification and operation authorization [6].

SSI utilizes three primary protocols: verifiable credentials, DIDs, and distributed ledger technology (DLT) [2]. DIDs and verifiable credentials are the main requirements for most blockchain-based identity systems [9]. In addition, the nature of decentralized infrastructure has shifted digital identification from an application to an ecosystem [10].

Verifiable Credentials: A VC is a digital credential that a person owns that contains information or qualities about them, for example, their name, date of birth, place of residence, and so on [6]. The verifiable credential is a standardized protocol by the World Wide Web Consortium (W3C), a fundamental SSI pillar. It ascertains that statements originating from a digital ID issuer are made in a manner that supports respect for privacy and is tamper-proof using public-key cryptography and digital watermarking [5]. Verifiable credentials ensure that the issuers' statements are made in a privacy-safe, tamper-evident manner [1]. The owner of a personal credential decides the components and quantity of digital identity to share with the verifier based on the request. The verifier verifies instant data without contacting the identity issuer [5]. Blockchain applications for SSI rely on two levels of authentication: selective disclosure and zero-knowledge proof to protect the privacy of the credential holder. The selective disclosure approach permits the generation of proofs for a credential based on specific attributes. The zero-knowledge proof uses cryptography, which aids in demonstrating a particular credential feature without disclosing its value. Users cannot prove their identity with verifiable credentials by revealing their personally identifiable information [1].

Decentralized Identifiers: DID is an essential component of SSI that enables the creation of private, secure, and distinctive peer-to-peer connections between two entities [1]. The W3C noted that DID is the "Uniform Resource Identifier," which is an identifier that can be used for credential exchange or authentication [9]. DID is the system that requires metadata to be published to the registry in the form of a new industry standard [10]. DID allows anyone to create their unique identifier to use in the digital world [6]. DIDs are intermediaries that use the central database to manage personal digital identity, where owners do not control how the data was collected or its usage [5]. Currently, users must rely on intermediary IDs to connect, such as those provided by email providers, *Facebook*, mobile network carriers, or *Google*. The intermediaries may combine the data with other metadata to deliver targeted advertisements; nevertheless, the data correlation techniques may have significant consequences. For example, two parties could exchange private DIDs to establish a safe route that is closed off to outsiders. Private DIDs enable the development of unique DIDs for various relationships to prevent data correlation [1]. Proof of ownership of the DID depends on possessing a private key connected to the public key included in the DID. The DID scheme is the formal syntax for a DID. At the same time, the DID method helps define the approach for implementing a particular system, where one can find information about the methods for creating, updating, and deactivating DIDs. The relevant data that describes the DID subject is available within the DID document; this includes public keys, other attributes, and metadata. The universal resolver provides the necessary features to address interoperability issues within decentralized identity management solutions [9].

Decentralized Identity: Decentralized identity models are made possible by decentralized technologies that eliminate the need for intermediaries. Traditional identity management systems rely on an IDP to store personal data centrally and manage it according to their own privacy and security policies. In such systems, both requesting parties and entities themselves must trust the IDP to ensure the availability, integrity, and confidentiality of their attributes. Decentralization aims to remove this single point of compromise by eliminating the central authority that serves as the IDP and is responsible for identity administration and management. Decentralized approaches rely on peer-to-peer relationships between interacting parties, rather than on centralized IDPs. This ecosystem comprises various entities that perform roles previously carried out by the central authority. Decentralized ledger technology (DLT), blockchain, distributed file systems, hash graphs, and tangles can all be used to implement decentralized identity with blockchain being the most popular option. Hyperledger Indy, an open-source project built explicitly for decentralized identity, can also be used alone or in conjunction with other blockchains [11].

Decentralized Key Management System: SSI Storage's overall objective is for users to use a personal portable device as a private repository or digital wallet for storing and managing credentials. Hence, a key recovery mechanism must be guaranteed in case of theft of digital wallets. There are two fundamental recovery approaches in a key management system: centralized and decentralized critical management systems. Centralized Key Management Systems (CKMS) enable users to store their private keys and credentials in a centralized key repository, such as cloud storage or offline backups, to recover the original credentials in case of loss or theft. These systems also provide a secure way for users to export their keys to hardware. In contrast, decentralized Key Management Systems (DKMS) use the *Shamir Secret Sharing* (SSS) protocol, which involves multiple entities, nodes, or individuals storing an individual's private key or essential seed [6].

Principles of SSI

The principles of SSI outlined by Naik and Jenkins provide a comprehensive framework for analyzing SSI solutions. These principles aim to ensure individuals have control over their personal data and online identity while also promoting transparency, fairness, privacy, and security. The following is a brief explanation of each principle.

Sovereignty: Individuals should have full autonomy and control over their identity without interference from external entities or authorities.

Existence of User: A digital identity can only be created for an individual who already exists in the real world and the purpose is to make certain aspects of the user publicly accessible.

Data Access Control: Identity owners should have the ability to access, update, share, conceal, or delete their personally identifiable information as per their discretion without interference from identity providers or authorized organizations.

Data Storage Control: Identity owners should have ownership and control over their personally identifiable information that should be stored in a system owned or controlled by the individual, rather than in centralized repositories or external control mechanisms.

Longevity: Individuals should have the ability to revoke or abandon their identity at any point while ensuring that the identity remains permanent as long as desired.

Decentralization: Identity management and registration should be decentralized, preferably utilizing public decentralized infrastructures like distributed ledger technology or decentralized networks.

Verifiability: Digital identities should have verifiable credentials, similar to physical credentials, and the verification process should not rely on direct interaction with the issuer.

Recovery: Reliable systems should be in place to restore an individual's identity in case of key, wallet, or device loss, ensuring resilience and continuity of the identity.

Cost-Free: Access to identity should be free of hidden costs or financial charges to ensure equal access for everyone, regardless of their ability to pay.

Security: Identity infrastructures should implement multiple layers of security, including secure connections, cryptographic measures, and decentralized and encrypted storage to protect identities and their associated communications.

Privacy: Identity infrastructure should minimize the collection of personal information and ensure that information is not tied to an individual's identity without their consent, preserving anonymity and privacy.

Safeguards: The rights and freedoms of identity owners should be upheld and independent authentication systems should prioritize the rights of identity owners in case of disputes.

Flexibility: Identity infrastructures should allow for flexibility in creating different types of identities, breaking down identity aspects, adding new features, and evolving identities over time.

Accessibility: Identity infrastructures and services should be user-friendly and accessible to a wide range of users, including vulnerable and non-technical individuals.

Availability: Equal access to identity infrastructure and services should be provided to everyone, regardless of their ethnicity, gender, socioeconomic status, or language without discrimination.

Transparency: Identity infrastructure should be based on freely available systems, protocols, and algorithms that adhere to open-source principles and are not tied to specific ownership or architecture.

Portability: Users should have the ability to easily transfer their identity and related data between different platforms, requiring standardized identity, credential, and data formats.

Interoperability: Different identity systems should be able to effectively communicate with each other on a large scale, enabling seamless communication between enterprises and government organizations.

Scalability: Identity infrastructures should have the capacity to handle the growing need for sovereign identity, efficiently accommodating the increasing number of digital entities.

Sustainability: Identity infrastructure and services should be sustainable in terms of the environment, economy, technology, and society in the long term.

By adhering to these principles, SSI solutions can empower individuals, enhance privacy and security, and establish a fair and transparent digital identity ecosystem.

SSI challenges

As discussed, SSI is a concept that allows individuals to have more control over their personal information and data. While the idea of SSI is appealing, the implementation is a complex problem and comes with many implications. The challenge of binding a physical person to their digital identity is a major concern and has led identity professionals to explore various ways to secure this relationship.

The hardest part of creating a digital identity is linking it to a real-life individual. There is a tendency for people to forget passwords and credentials, which creates security risks. In addition, the loss of bitcoins due to forgotten passwords and damaged hard drives illustrates the dangers of entrusting sensitive information to fallible individuals.

Having an infrastructure that relies solely on people's memories to determine ownership of assets is a questionable approach, as it increases the risk of loss and the inability to recover assets. In an SSI system, where each user holds a private key, there are no backup mechanisms in place to help recover a lost or forgotten key. The security of the key, protected by complex algorithms, makes it nearly impossible to brute-force an attack but also means that there is no way for anyone to assist if a key is lost or forgotten.

The responsibility of keeping track of and securely storing the private key falls solely on the user and there is no one to turn to for help if it is lost. This highlights the importance of creating reliable and secure recovery mechanisms that can help individuals regain access to their digital identity in the event of a lost key. Without proper recovery options, the system is vulnerable to disaster and loss, as there is no one to call for assistance.

It is evident that there is a need for identity custodians, a trustworthy entity that can be relied on in case of any problems and can provide access to a user's private key if it is lost. In order to prevent rogue employees of the custodian from gaining access to the private key, high-security measures must be implemented, but at the same time, the user needs to be involved in the recovery process. However, this creates a challenge, as high-security measures and ease of use are not always compatible.

The other challenge is proving identity when the private key is lost, and no other possessions can be used. This requires a secure and efficient key recovery system that is easy to use. Although no system can be 100% secure, it is crucial to have a high level of security, given the importance of keeping private keys private.

According to *Opinion Piece*, splitting the key into multiple parts and storing them physically can make the system more secure against digital attacks. To ensure a biometric match, the user must be physically present to collect all the parts and reconstruct the private key. The identity custodian must follow proper procedures to ensure the privacy of the user's identity.

Creating a secure, cost-effective, and usable system for managing identities is not easy. SSI, which is often seen as a straightforward solution, actually requires complicated approaches and the participation of multiple custodians [12].

Potential use cases of SSI in blockchains

As reported by *Pande* in 2022, the Ethereum blockchain has introduced SSI systems that allow for password-free access throughout the ecosystem. With SSI, users are identified through their unique proxy profiles, including humans, groups, organizations, and bots. The system provides a secure, transparent, auditable, and error-free solution for identity management, benefiting both beneficiaries and service providers [13]. Additionally, the following benefits can be seen in various areas of human and system endeavors as outlined below.

> *Data Privacy in Health Care*: The confidentiality of medical records is of utmost importance in the healthcare industry. To ensure data privacy, *Lumedic Exchange* utilizes SSI-based authentication to securely transmit healthcare records between insurers, hospitals, and pharmacies [13]. Lumedic Exchange implements a health information exchange model that prioritizes patients and eliminates the requirement for centralized storage of confidential data. Instead, patients are given control and ownership over their information, granting access to sensitive health details only when necessary for specific health-related purposes [14].
>
> *Increasingly Efficient Financial Services*: Financial institutions can benefit greatly from SSI systems that enhance the speed of SSI. Bonifil's *MemberPass* is an SSI-based identity authentication system that solves problems related to credit union services in the United States [13]. As captured by Jennifer Land, MemberPass helped credit unions to prevent fraud before it happens, complete member enrollment with minimal friction, ensure consistent authentication experiences across all channels and transactions, eliminate passwords and challenge questions, avoid hackable central databases of personally identifiable information, and reduce expenses due to better fraud protection and shorter authentication times [15].
>
> *Staff Identification System*: SSI offers budget-friendly employee identification solutions that enhance the safety and mobility of emergency health workers in the event of a crisis. By providing cost-effective solutions for employee identification, SSI allows for unrestricted movement during emergencies, giving doctors, nurses, and other healthcare professionals the ability to quickly and easily respond to a crisis [13]. This not only improves the efficiency of emergency response efforts but also helps ensure the safety of both the health workers and those in need of assistance. When responding to an emergency, every second counts, and SSI's solutions give emergency responders the freedom and flexibility to respond quickly and accurately.
>
> *Competency Assurance*: In many parts of the world, the need for professional identity verification is a pressing issue, as there are frequent instances of people falsely claiming qualifications they do not possess. When discussing the role of blockchain in establishing SSI, another important area to consider is that of human resources and ensuring the competence of individuals. Ensuring competence is especially important in fields such as medicine, where a doctor must be properly trained to perform surgery, or in the aviation industry, where pilots must be trained and be medically fit to operate certain aircraft. It is important to ensure that these professionals have the necessary certifications and qualifications to perform their jobs, and this is where blockchain-based SSI comes into play. A person's ability to perform a specific task can be verified using SSI, particularly in situations where errors could have fatal consequences [16].

Know Your Customers: The use of SSI has numerous applications, and one of the most well-known within the banking industry is *Know Your Customer (KYC)*. Many financial processes require user identity verification, such as transactions that need payment. SSI enables a more streamlined way of ID verification through a reusable KYC concept. When identity verification is needed, SSI can significantly reduce the hassle for users and improve the customer experience while providing a compliant service. In other words, traditional KYC is single-use, while SSI makes KYC reusable. The reduction of friction is not limited to just Business-to-Consumer (B2C) interactions, it also applies to Business-to-Business (B2B) scenarios. SSI also provides traceable and auditable personally identifiable information (PII). Additionally, its cross-industry applicability accelerates the onboarding process for new customers and improves security by eliminating the need for paper-based processes. SSI is therefore an ideal solution for enabling mobile banking. In a recent trial, the UK government used SSI technology to onboard users for its *Financial Conduct Authority* (FCA) [17].

Authentication, Authorization, and Trust of IIoT User/Devices: There are several organizations, both public and private, that have implemented identity management solutions to control authentication and authorization within and across systems. These solutions use physical devices like chip cards, knowledge-based methods requiring passwords and tokens, and inherent biometric features. These solutions usually rely on centralized authentication and identity network services, which allow users to authenticate themselves using digital credentials from third parties. Most of these solutions are also applicable to the Industrial Internet of Things (IIoT) and provide security through asymmetric cryptography, but this can lead to high computational costs. Among the options for addressing these limitations are relying on third parties or blockchains or preventing data exposure. Another solution is to use SSI to ensure privacy without relying on third parties. For instance, the Sovrin Foundation offers a public utility for SSI on the internet. Sovrin is a protocol and token for SSI and decentralized trust, and uses self-sovereign and verifiable decentralized digital identifiers (DIDs) to authenticate identities, preserving privacy through its network design [18].

Part Lifecycle Support: The lifespan of industrial automation systems tends to be very long. There are currently assembly lines and facilities in the process industry that have been operating continuously for over 20 years, and this trend is expected to continue with the rise of the IIoT. Over these extended life cycles, the parts, components, and machines in future factories will need to be maintained, replaced, or reconfigured. As a result, it is crucial to guarantee that new parts, whether from original equipment manufacturers or other sources, are genuine, safe, and fully compatible with existing ones. SSI technologies provide a secure, decentralized method for ensuring the authenticity of components, parts, and machines throughout the life cycle, making them ideally suited for managing the life cycle of parts, components, and machines. Additionally, the long-term life cycle of parts and components can be recorded and verified to demonstrate that they were made by a specific brand or possess specific attributes [18].

Non-Fungible Tokens (NFT): SSI plays a crucial role in verifying the creation, ownership, and current possession of NFTs throughout their life cycle, as well as providing proof of ownership of partial NFTs. It eliminates the challenge of proving NFTs' origin, regardless of the ledger on which they are stored. This enables a decentralized content consumption system with integrated payment and identity, allowing for direct interaction between content creators and their audience, thus ensuring fair payment for their work. Additionally, SSI addresses the issue of identity linked to payments in cryptocurrency and decentralized finance (DeFi) transfers, where the identity of payment recipients is not easily verifiable beyond their wallet address. Particularly when it comes to high-value transfers, identity verification has become increasingly important. The anonymous nature of cryptocurrency payments remains a requirement, but there are times when confirming the recipient's identity is advantageous [17].

CONCLUSIONS AND RECOMMENDATIONS

This chapter has provided an exploration of the evolution of digital identity, from centralized identity management to SSI management. The key findings and recommendations are as follows:

- SSI systems offer a secure and resilient solution for identity management. Organizations should consider implementing SSI systems to benefit from increased security, privacy, and control over personal data.
- SSI distinguishes itself from decentralized identity by emphasizing individual control and ownership of personal data. It utilizes DIDs and key management systems to ensure secure and verifiable identities.
- Two main methods for key recovery in SSI systems are CKMS and DKMS. Organizations should assess their needs and choose the appropriate method based on factors such as security, hardware exporting, and distribution across multiple entities.
- The 20 extended sets of principles proposed by Naik and Jenkins serve as a foundation for SSI. These principles prioritize individual sovereignty, data control, security, privacy, and interoperability, among other essential aspects of identity management (Table 7.2).
- SSI systems have diverse use cases, including healthcare, financial services, staff identification, KYC processes, and NFTs. Organizations in these sectors can leverage SSI to enhance security, streamline processes, and improve customer experience.

Recommended good practices for SSI systems

- Build a secure peer-to-peer digital network between the ID issuer, verifier, and owner, leveraging verifiable credentials, DIDs, and distributed ledger technology.
- Utilize digital wallets owned and controlled by the user on their devices, ensuring user-centric control and security of personal data.
- Implement PKI systems using digital wallets and distributed ledgers for end-to-end interaction among users, including issuance, update, certificate application, and revocation.
- Employ Decentralized Key Management Systems (DKMS) with the SSS protocol to store private keys across multiple entities or nodes, enhancing security and resilience.

Table 7.2 Principles of self-sovereign identity

Sovereignty	Existence of user	Data access control
Data Storage Control	Longevity	Decentralized
Verifiability	Recovery	Cost Free
Security	Privacy	Safeguard
Flexibility	Accessibility	Availability
Transparency	Portability	Interoperability
Scalability	Sustainability	

- Split the private key into multiple parts and store them physically to provide additional security against theft or loss of digital wallets.
- Establish a key recovery mechanism in case of digital wallet theft or loss.
- Adopt an effective governance model that defines policies, principles, standards, and terminology, ensuring transparency, accountability, and trust among all participants.
- Utilize selective disclosure and zero-knowledge proof techniques to protect privacy and ensure data confidentiality within the SSI system.

By implementing these recommended practices, organizations can develop a secure, decentralized, and user-centric SSI system. These measures enhance security, privacy, interoperability, and trust among all participants, ultimately improving the overall effectiveness and usability of the system.

REFERENCES

[1] Geroni, D. (2022, August 15). *Self-sovereign identity: The Ultimate Beginners Guide!* 101 Blockchains. Retrieved November 19, 2022, from https://101blockchains.com/self-sovereign-identity/

[2] T. Hahn, "The Evolution of Digital Identity," helix id, Mar. 27, 2020. https://medium.com/helix-id/the-evolution-of-digital-identity-cc54917a6a49 (accessed Oct. 26, 2022).

[3] Preukschat, A., Reed, D., & Searls, D. (2021). *Self-sovereign identity: Decentralized Digital Identity and verifiable credentials*. Manning.

[4] Allen, C. (2016, April 25). *Life with alacrity*. The Path to Self-Sovereign Identity. Retrieved February 12, 2023, from http://www.lifewithalacrity.com/2016/04/the-path-to-self-sovereign-identity.html

[5] Okta Updated: 07/21/2022 - 11:19 Time to read: 9 , & Okta. (2022, July 21). *Self-Sovereign Identity (SSI): Autonomous Identity Management*. Okta. Retrieved November 19, 2022, from https://www.okta.com/identity-101/self-sovereign-identity/

[6] Bai, Y., Lei, H., Li, S., Gao, H., Li, J., & Li, L. (2022). Decentralized and self-sovereign identity in the era of Blockchain: A survey. *2022 IEEE International Conference on Blockchain (Blockchain)*, 500–507. https://doi.org/10.1109/blockchain55522.2022.00077

[7] N. Naik and P. Jenkins, "Governing principles of self-sovereign identity applied to blockchain enabled privacy preserving identity management systems," *IEEE Xplore*, 2020. [Online]. Available: https://ieeexplore.ieee.org/document/9272212. [Accessed: 27-Oct-2022].

[8] Mahula, S., Tan, E., & Crompvoets, J. (2021). With blockchain or not? opportunities and challenges of self-sovereign identity implementation in public administration. *DG.O2021: The 22nd Annual International Conference on Digital Government Research*, 495–504. https://doi.org/10.1145/3463677.3463705

[9] Weston, G. (2022, September 8). *Self Sovereign Identity & Decentralized Identity - an unlimited guide*. 101 Blockchains. Retrieved November 19, 2022, from https://101blockchains.com/self-sovereign-identity-and-decentralized-identity/

[10] Baya, V. (2019, October 9). *Digital Identity: Moving to a decentralized future*. Online Banking, Mortgages, Personal Loans, Investing. Retrieved November 20, 2022, from https://www.citi.com/ventures/perspectives/opinion/digital-identity.html#:~:text=Digital%20identity%20has%20evolved%20from%20centralized%20to%20federated,providers%2C%20this%20solution%20is%20neither%20secure%20nor%20efficient

[11] Čucko, Š., & Turkanovic, M. (2021, October 4). *Decentralized and self-sovereign identity: Systematic mapping study*. IEEE Xplore. Retrieved November 20, 2022, from https://ieeexplore.ieee.org/document/9558805

[12] Piece, O. (2018, September 10). *The problem with self-sovereign identity: We can't trust people*. Payments Cards & Mobile. Retrieved November 20, 2022, from https://www.paymentscardsandmobile.com/the-problem-with-self-sovereign-identity/

[13] Pande, S. (2022, October 6). *Self-sovereign identity (SSI) benefits in the security of Digital Identity*. Latest Crypto, Blockchain, Metaverse and Web3 News. Retrieved November 20, 2022, from https://blockchainshiksha.com/self-sovereign-identity/

[14] Barry, Q. (2021, June 25). *The lumedic exchange aims to modernize data privacy in healthcare - identity review*. Identity Review | Global Tech Think Tank. Retrieved February 4, 2023, from https://identityreview.com/lumedic-exchange-modernizes-data-privacy-in-healthcare/

[15] Land, J. (2022, August 23). *Next generation MemberPass delivers multiple new functions and attractive user benefits*. Bonifii. Retrieved February 4, 2023, from https://bonifii.com/2022/08/next-generation-memberpass-delivers-multiple-new-functions-and-attractive-user-benefits/

[16] Louw, L. (2022, August 10). *Self-sovereign identity on the blockchain - use cases*. BSV Blockchain. Retrieved November 23, 2022, from https://bsvblockchain.org/news/self-sovereign-identity-on-the-blockchain-use-cases/

[17] Team Blockchain. (2021, October 15). *Self-sovereign identity use cases*. DLA Piper Investment Rules of the World. Retrieved February 12, 2023, from https://www.dlapiperintelligence.com/investmentrules/blog/articles/2021/self-sovereign-identity-use-cases.html

[18] Bartolomeu, P. C., Vieira, E., Hosseini, S. M., & Ferreira, J. (2019, October 17). *Self-sovereign identity: Use-cases, technologies, and ... - IEEE Xplore*. 2019 24th IEEE International Conference on Emerging Technologies and Factory Automation (ETFA). Retrieved February 12, 2023, from https://ieeexplore.ieee.org/document/8869262/

Chapter 8

Enhancing voting/e-voting through blockchain

Alamgir Faruque and Tahziba Tabassum

BACKGROUND

Elections worldwide have been marred by allegations of corruption, vote manipulation, and abuse of power, resulting in a decline in public trust. The declining voter turnout in countries like Iran, Russia, and the United States has raised concerns about the legitimacy of elected governments. A critical point of discussion in this context is the choice between paper-based and electronic voting methods. Past experiences often influence opinions on this matter. However, the utilization of blockchain technology in electronic voting holds the potential to revolutionize the situation by offering transparency and security, thereby restoring voter confidence in the electoral process.

Former US President Donald Trump's legal challenges against the 2020 US presidential election results were dismissed; however, a significant portion of his supporters continued to believe in fraudulence. A survey conducted in December 2020 revealed that over 75% of Republican voters claimed millions of fraudulent ballots were cast and thousands of deceased individuals' votes were registered [1]. Similarly, in Russia, the credibility of election results has been questioned, with many citizens alleging fraud. *Golos Russia*, a reputable election monitoring organization, has exposed severe violations of electoral integrity, reinforcing suspicions of compromised elections. These violations, which occurred on the eve of the presidential elections in March 2012, were more evident than interference with ballots on election day [2]. The low voter turnout during the 2021 Iranian elections also signals the persisting grievances of the people against the authoritarian government [3].

Numerous studies conducted in the past decade in the United States and internationally have emphasized the significance of voting experiences in shaping voter trust in the political process. Interactions with voting mechanisms, exposure to party language, and media reports can create a distorted perception of the risk of electoral fraud, further eroding trust. Addressing these issues is crucial to safeguard the integrity of the electoral process and promote trust among voters. Blockchain technology, with its decentralized and tamper-proof nature, offers a potential solution to enhance transparency and security in voting systems. Furthermore, implementing voter identity verification through biometric technologies can further mitigate voter fraud. Transparency and voter trust are vital for the stability of democratic systems [4].

Freedom House, an organization that advocates for democracy and defends democratic values, has identified countries with deteriorating scores for political rights and civil liberties. According to their annual *Freedom in the World* report for 2022, the number of countries experiencing declines in aggregate scores has exceeded those with gains for the past 16 years [5]. Freedom House has identified 16 countries with the lowest aggregate scores, indicating limited democratic practices. However, some countries with higher scores demonstrate relatively better democratic progress (Table 8.1).

DOI: 10.1201/9781003462033-10

Table 8.1 Scoring for countries with the worst scores for political rights and civil liberties [5]

Country	Aggregate Score
South Sudan	1
Syria	1
Turkmenistan	2
Eritrea	3
North Korea	3
Equatorial Guinea	5
Central African Republic	7
Saudi Arabia	7
Somalia	7
Belarus	8
Tajikistan	8
Azerbaijan	9
China	9
Libya	9
Myanmar	9
Yemen	9

Russia: suppression of political opposition

Under the leadership of President Vladimir Putin, the regime in Russia has intensified its efforts to suppress political opponents and civil society groups, effectively eliminating any meaningful competition in the parliamentary elections held in September. These actions have raised concerns about the lack of political pluralism and democratic practices in the country.

Iran: shifting political dynamics

In Iran, the parliamentary elections witnessed a decline in the political influence of Iran and its associated militia groups. Parties aligned with these groups were defeated, resulting in a political landscape with fewer irregularities compared to previous elections. However, the overall democratic situation in Iran remains a subject of scrutiny.

Myanmar: coup and repression

Following the defeat of their preferred political party, the military in Myanmar seized control through a coup, preventing the newly elected parliament from convening. Subsequently, they resorted to violent suppression of the pro-democracy protest movement, using deadly force. This has led to significant concerns regarding the erosion of democratic values and human rights in the country [5].

United States: Trump's influence and republican party dynamics

In the United States, former President Donald Trump garnered an unprecedented 74 million votes in the 2020 election, despite running a disorganized campaign. His influence within the Republican Party has solidified, with his base of supporters, known as the *MAGA Movement* considering him a powerful and unpredictable political force. Trump's second impeachment and the subsequent media attention surrounding the Senate trial have further fueled beliefs among his supporters that he was a victim of a rigged election or global elite conspiracy. Some Republicans in Congress, including senators and representatives, contested the final Electoral College results and have not been deterred by the events of January 6th.

The Trump Republicans, driven by an unapologetic narrative, are even exploring the formation of a new political party called *The Patriot Party* and supporting primary challenges against current Republican Party members. This determined bloc of Trump supporters has positioned themselves as a militant opposition, similar to historical movements like the Anti-Federalists or Southern Democrats. This dynamic within the Republican Party may present a potential threat to American democracy [6].

To drive the point home, various countries face distinct challenges to their democratic systems. Russia's suppression of political opposition raises concerns about the lack of political pluralism, while Iran experiences shifting political dynamics. Myanmar's military coup and subsequent repression have sparked fears of human rights violations. In the United States, the continued influence of former President Trump and his dedicated supporters within the Republican Party poses challenges to democratic norms. These ongoing developments highlight the importance of safeguarding democratic principles and fostering an inclusive political environment to ensure the integrity of democratic processes worldwide.

DISCUSSION

Blockchain technology, with its decentralized and immutable ledger, holds significant potential for transforming the voting process. Unlike traditional centralized voting systems, blockchain-based systems offer enhanced security and transparency. The decentralized nature of blockchain makes it highly resistant to hacking attempts, as modifying data would require controlling the majority of nodes, which is practically impossible. This feature ensures the integrity of the voting records and reduces the likelihood of fraud or manipulation.

In contrast, traditional voting systems that rely on centralized databases are vulnerable to cybersecurity incidents. A single server failure can bring down the entire system, compromising the integrity of the votes. However, with blockchain-based voting systems, even if a single node is compromised, the network remains operational, as the consensus mechanism requires a majority to be compromised.

Identity theft, a common concern in centralized voting systems, is mitigated by blockchain technology. Users' identities need to be validated before they can participate in the voting process, reducing the risk of impersonation. The use of cryptographic encryption and private and public key combinations adds an extra layer of security, making it difficult for external actors to interfere with the process.

The transparency and accuracy provided by blockchain technology are also significant advantages for voting systems. Each vote is recorded and counted accurately, leaving an immutable trail that prevents manipulation. The technology's cryptographic techniques protect voter confidentiality while still ensuring transparency in the overall process.

Another notable benefit of blockchain-based voting systems is their potential to reduce costs and environmental impact. Traditional voting methods, such as paper ballots, consume substantial resources and generate waste. In contrast, blockchain-based systems are digital and do not require physical resources, saving money and reducing environmental harm. Online voting also increases accessibility and convenience, enabling individuals who are unable to physically visit polling stations to participate.

However, it is essential to acknowledge the challenges associated with blockchain technology in voting systems. The environmental impact of cryptocurrency mining, which still supports some blockchain networks, is a concern. However, combining a large amount of electronic information into one single code (called a "transaction hash value") can reduce the number of codes needed. This reduction helps make blockchain-based voting systems more environmentally friendly by decreasing their impact on carbon emissions.

To recap, blockchain technology offers a promising solution for improving the voting process, providing transparency, security, and accuracy. Its potential to reduce costs, increase accessibility, and mitigate environmental impact makes it an attractive option for governments and organizations. While addressing the environmental impact of cryptocurrency mining remains a challenge, the benefits of blockchain technology in voting systems are substantial [7].

Constraints of traditional voting systems

It is important to carefully address the following constraints and challenges to maximize the potential benefits of blockchain technology in voting systems, such as increased security, transparency, accessibility, and efficiency.

Security Vulnerabilities: Traditional voting systems, particularly electronic voting systems, are susceptible to hacking and manipulation. The complex software and hardware used in these systems can be exploited by malicious actors, compromising the integrity of the voting process.

Lack of Transparency: Existing voting systems often lack transparency, making it difficult to detect and prevent fraud. The centralized nature of these systems can result in a lack of accountability and limited visibility into the entire voting process.

Limited Accessibility: Traditional voting systems may pose challenges for certain groups of people, such as those with disabilities or citizens living in remote locations. Physical polling stations may be inaccessible or impractical for some individuals, leading to lower voter participation.

Cost and Inefficiency: Conducting elections through traditional methods, such as paper ballots, can be expensive and time-consuming. The process of collecting, transporting, and counting physical ballots can be labor-intensive and prone to errors.

Lack of Anonymity: Some voting systems compromise voter anonymity, which can result in coercion and intimidation. Maintaining the privacy of individuals' votes is crucial to ensure a free and fair electoral process.

Environmental Impact: Traditional voting systems that rely on paper-based methods consume significant resources, including paper and energy. This can have a negative environmental impact, contributing to deforestation and carbon emissions.

Real-life use cases of blockchain e-voting

The use of blockchain technology in e-voting has been implemented in several real-life cases, including the state of West Virginia in the United States and the Victorian Electoral Commission (VEC) in Australia.

In West Virginia, the state became the first in the United States to introduce blockchain-based internet voting in its primary elections. The system was developed by *Voatz*, a Boston-based startup, and utilized blockchain technology to secure and validate votes. Voters were able to cast their ballots using a mobile app, which incorporated facial recognition and biometric verification to ensure that the person voting was a registered voter. The votes were then recorded on a distributed ledger, or blockchain, making it extremely difficult to tamper with or compromise the results. The system has been used in multiple elections in West Virginia since 2018, although concerns about security and dependability of the technology have been raised. Other states in the United States have also considered adopting blockchain for voter registration and ballot tracking, although West Virginia remains the only state to implement blockchain-based voting as of this writing.

The West Virginia election using blockchain technology served as a pilot project to test the technology, without immediate plans for large-scale implementation. However, the potential benefits of using blockchain in elections extend beyond this experiment. By providing a secure and tested interface for mobile voting, blockchain technology could help reduce voter fraud and increase voter turnout. It could also make voting more convenient for citizens, including those who are abroad. Additionally, the use of blockchain technology could benefit election commissions by enhancing transparency, reducing costs, streamlining the vote-counting process, and ensuring the accuracy of all votes [8].

As mentioned before, in the West Virginia elections, voters were required to verify their identity through biometric tools such as thumbprint scans before casting their vote on a mobile device. Each vote was then recorded on a chain of votes that was mathematically proven by a third-party participant. The use of blockchain technology ensured that all election data was publicly recorded on a verifiable ledger while maintaining the anonymity of voters. The results were also available immediately.

To further enhance the use of blockchain in elections, open-source blockchain voting platforms can be utilized. These platforms do not contain proprietary elements allowing citizens and agencies to review and improve the security of the application. Open-source systems play a crucial role in ensuring secure and reliable elections. Several startups, such as *Democracy Earth Foundation, Follow My Vote, Democracyos.org, VoteWatcher, Milvum,* and *VotoSocial,* are developing innovative technologies based on the open data philosophy [9].

The Victorian Electoral Commission (VEC) in Australia has also improved its e-voting system with a cryptographic e-voting system that offers better security and verifiability. The system provides both individual verifiability, allowing voters to confirm that their vote has been recorded correctly, and universal verifiability, which provides mathematical proof that all votes have been accurately recorded and tallied. Verifiability is essential for ensuring that the election has not been manipulated or contains errors. However, it is important to note that the level of verifiability provided by a system may be challenging for non-experts to understand. Some e-voting systems, such as those used in New South Wales and Victoria, use a receipt code mechanism that confirms the storage of receipt codes but not the vote integrity. Verifiability must be incorporated into the e-voting system from the start and cannot be added later without compromising privacy.

The Victorian Electoral Commission (VEC) in Australia is actively working on developing a new e-voting system that aims to replace its current vendor-provided machines. The purpose of this new system is to assist voters with disabilities and potentially cater to up to 400,000 citizens. The VEC is collaborating with research partners to adapt the *Prêt à Voter* (ready for vote) system for Alternative Vote (AV) and Single Transferable Vote (STV) preferential voting, with the intention to deploy it on electronic devices.

The Prêt à Voter system offers both individual verifiability, allowing voters to detect any tampering with their votes and universal verifiability, which enables anyone to verify the election results. This cryptographic verifiability provides strong assurance against manipulation of votes and ensures the integrity of the election process. The VEC aims to maintain these verifiability characteristics for use in statewide elections, ensuring transparency and trust in the system.

One of the key challenges the VEC faces in developing the new e-voting system is striking a balance between security and usability, particularly for voters with disabilities. Cryptographic voting systems may involve additional steps for verifying cryptographic operations, which could potentially pose challenges for some voters. The VEC acknowledges this challenge and is actively investigating various cryptographic designs that offer different levels of security and usability trade-offs. Prototypes are being tested to evaluate and enhance the designs based on user feedback and requirements.

In addition to the new e-voting system, the VEC is also working on a remote e-voting system to provide private and secure voting options for specific groups, such as vision-impaired voters who cannot use postal voting. The VEC is collaborating with e-voting experts to create a more secure system, and the code will be made available under an open-source license, allowing other electoral commissions to utilize it as well. This approach promotes transparency and collaboration in improving e-voting systems.

However, it's important to note that remote e-voting, while providing convenience, poses certain security risks, especially regarding ensuring a secret ballot. Security experts have expressed concerns about the elimination of the ability to guarantee vote privacy in remote e-voting systems. Currently, remote electronic voting (REV) is not permitted under the Victorian Electoral Act. The VEC has proposed REV to the parliament for consideration specifically for vision-impaired voters, and they are evaluating several peer-reviewed REV systems, including *Helios 6* and *Civitas 7*, to address the associated risks and challenges.

The VEC's efforts in developing these e-voting systems demonstrate their commitment to improving accessibility, security, and transparency in the electoral process. Through collaborations, testing, and careful consideration of trade-offs, they aim to create a reliable and inclusive voting system that upholds the integrity of elections in Victoria, Australia.

Remote electronic voting (REV) conducted through direct access introduces significant risks for individual voters and the overall election process. Evaluating REV protocols while considering the trade-offs between these risks and the right to vote is essential. Additionally, the privacy of the two-envelope postal voting system relies on adherence to proper procedures by the electoral commission. However, a 1992 court case determined that this procedure is not legally considered secret. Internet voting presents even greater risks, emphasizing the need for cautious deployment and limited scope for REV systems.

To mitigate the risks associated with Internet voting, the Victorian Electoral Commission (VEC) plans to limit its use to 5,000 voters out of a total of 3,200,000 voters. Furthermore, a rigorous registration and authentication process will be implemented to enhance security measures. A notable example is the approach taken by Norwegian authorities in 2011, where they emphasized the importance of openness and verifiability in their Internet voting system. By making the system open, a thorough evaluation of its verifiability was conducted before deployment, leading to potential enhancements to the protocol [10, 11].

These real-life use cases of blockchain in e-voting demonstrate the potential of the technology to enhance security, transparency, and verifiability in the electoral process. However, challenges remain, and further research and testing are necessary to ensure the feasibility and effectiveness of blockchain-based voting systems.

Challenges of blockchain implementation in e-voting

Despite its many potential benefits, a blockchain-based voting platform may also present a number of implementation challenges as follows:

Technology Readiness: Blockchain technology is still relatively new and untested for large-scale use in elections. Further research and development are needed to ensure its reliability, scalability, and security.

Comprehensive Understanding: Government officials and public administrators must have a thorough understanding of blockchain technology and its intricacies to effectively implement it in voting systems. Stakeholder input and collaboration are crucial for successful deployment.

Cost-effectiveness and Scalability: Blockchain solutions need to be reasonably priced and scalable to accommodate a larger user base. Ensuring cost-effectiveness is essential to gain support for adopting blockchain-based voting systems.

Usability and Accessibility: Balancing security and usability is crucial in designing blockchain-based voting systems, especially for voters with disabilities. User-friendly interfaces and clear instructions are important to ensure inclusivity and accessibility.

Political Support: Adoption of new technologies in the electoral process requires political support and willingness to embrace innovative solutions. Building awareness among government officials and gaining their support is crucial for the successful implementation of blockchain-based voting systems [12].

On-going research

In 2019, a research paper [11] proposed the utilization of blockchain technology to bolster the security of electronic voting systems. The suggested system encompasses key designs, including a synchronized model of voting records based on Distributed Ledger Technology (DLT) to prevent vote manipulation, a user credential model employing elliptic curve cryptography (ECC) to ensure authentication and non-repudiation, and a withdrawal model enabling voters to modify their vote before a specific deadline. By integrating these designs, the blockchain-based e-voting scheme addresses the essential requirements for electronic voting.

The proposed system operates as a public, distributed, and decentralized platform, enabling voting from diverse mobile devices and computers. It not only offers voters an economical means to audit and verify their votes but also ensures the security of their data through the use of blockchain technology. The voting blockchain acts as an expanding list of voting blocks, with voters casting their ballots and the voting office verifying and querying the votes. A distributed server of timestamps on a peer-to-peer network manages the vote database while the public key infrastructure (PKI) handles public-key encryption. The vote database tracks voting statistics and is updated by the voting office, while miners are responsible for validating and adding accepted votes to the public voting blockchain.

As discussed, the primary objective of the proposed blockchain-based electronic voting system is to enhance the security and usability of electronic voting. Through the incorporation of designs such as a synchronized model of voting records, a user credential model, and a withdrawal model, the system successfully meets the critical requirements for electronic voting. Furthermore, its public, distributed, and decentralized nature facilitates the recording and verification of votes while safeguarding the integrity of voters' data [11].

Decentralized e-voting on the ethereum blockchain

A research paper titled *Decentralized Voting Platform Based on Ethereum Blockchain* introduces a decentralized online voting system built on blockchain technology. The proposed system leverages smart contracts to ensure the security, transparency, and tamper-proof nature of the voting process. By addressing issues encountered in traditional voting systems, such as voter fraud, vote manipulation, and centralized control, the platform offers increased transparency, security, and trustworthiness while minimizing errors and malicious activity. The paper presents a proof-of-concept implementation, highlighting the potential of this decentralized voting platform for national and regional elections, as well as other voting applications. Although still in development, it represents a significant advancement toward secure and transparent online voting using blockchain technology.

Traditional e-voting systems, based on a client-server architecture controlled by a central authority, face challenges including insecure data, single points of failure, centralized control, and uncertain runtime environments. In contrast, blockchain technology provides decentralization and ensures data immutability through cryptographic functions and consensus algorithms.

The Ethereum Blockchain serves as a publicly accessible platform for creating Decentralized Applications (DApps) that offer features like a secure runtime environment, decentralized control and verification, strong data integrity, publicly accessible business rules, and high availability. The proposed solution offers several advantages compared to current electronic voting systems. It ensures that voting data is secure and cannot be altered, guarantees the reliability and stability of the system, and allows for decentralized registration and verification of voters. Additionally, it promotes transparency and clarity in the voting process by making smart contract votes publicly visible. The solution also restricts each voter to casting only one vote per Mobile Station International Subscriber Directory Number (MSISDN) and safeguards the privacy of recorded votes.

The proposed decentralized voting system consists of several key components: a web application, an event management server, smart contracts, an SMS gateway, and a mobile application. Two types of smart contracts are employed: a registration contract for secure user registration and authentication, and multiple voting contracts each deployed for a specific voting event with distinct questions and answers. The SMS gateway plays a crucial role in user authentication by sending SMS messages to users' phone numbers. Voters utilize the mobile application to register, cast their votes, and view real-time voting results, connecting to the blockchain network through an integrated Ethereum light client. To become an eligible voter, users must register with the system through the mobile application. The application retrieves the user's phone number from their SIM card and generates an empty Ethereum wallet, which users must fill with sufficient Ether balance. A smart contract facilitates a secure and automated registration process. After users input their information, the contract generates a unique Personal Identification Number (PIN) code and sends it to the user's phone via an SMS message. Users then enter the PIN into the application to complete the registration process. Using the web application, event organizers input the necessary information to create a voting event. The smart contract ensures that duplicate votes are rejected, allowing each phone number to vote only once. Organizers can monitor real-time updates and visual representations of ongoing event statistics.

The proposed platform's effectiveness was evaluated through the utilization of various technologies. *Solidity* was employed to write the smart contracts, while NodeJS facilitated server-side scripting in the Event Management Server. Web3js was used to interface with the Ethereum light client, and the mobile application was developed using HTML5 and compiled with *Apache Cordova*. The *Ropsten Testnet* simulated the Ethereum network, and Twilio served as the SMS gateway API. The user registration process typically took two to four minutes, mainly during the initial setup, while the act of voting required approximately 40 seconds to two minutes per vote [13].

CONCLUSIONS AND RECOMMENDATIONS

Advancements in technology have significantly improved various aspects of human life, including voting and e-voting systems. However, the security and integrity of e-voting systems remain a critical concern. Confidentiality, integrity, and availability are three key aspects that must be ensured to maintain the effectiveness of e-voting systems.

In recent years, several countries have experienced a decline in voter participation, primarily due to a lack of integrity in the voting process. This erosion of trust has been further compounded by the rise of extremist ideologies as seen in events like the attack on the US Capitol Hill. Such incidents underscore the urgent need to strengthen the integrity of the electoral process and address the underlying issues that have contributed to the rise of extremist ideologies.

Blockchain technology has emerged as a potential solution to enhance the security and integrity of e-voting systems. Its decentralized and transparent nature makes it difficult for malicious actors to manipulate the voting process. By incorporating a biometric authentication system and utilizing a shared, immutable ledger, e-voting systems can be made more secure. However, it is important to consider certain factors when implementing blockchain technology in e-voting systems. Processing times are typically longer compared to other methods, and the high computational power required for blockchain operations can result in increased electricity consumption. These issues need to be addressed to ensure the efficiency and sustainability of blockchain-based e-voting systems.

In short, continued research and development in the field of e-voting technology, particularly with the integration of blockchain, hold promise for improving the security, transparency, and trustworthiness of voting systems. It is essential to prioritize the enhancement of confidentiality, integrity, and availability in e-voting systems to safeguard the democratic process and restore public trust in elections.

Best practices for e-voting

Based on the above discussion the following are suggested as best practices that can help any private or government entity to implement a transparent e-voting framework using blockchain technology.

Authentication Methods: Voter authentication is a critical aspect of e-voting as it ensures the identity of the voter in a remote environment. The traditional approach involves providing a username and password to the voter during the registration phase, which is then required at the time of casting the vote. However, this approach is not secure as the stored credentials in the voting server can be stolen or modified by external attackers to impersonate valid voters. They are also vulnerable to eavesdropping attacks that intercept the submitted passwords. To address these security concerns, strong authentication methods such as one-time passwords or digital certificates can be used. One-time passwords change each time the voter authenticates making the intercepted credentials useless. Digital certificates provide the most robust solution as they provide access and data authentication by digitally signing the vote. The vote is encrypted before being signed to prevent the digital signature from being used to link voters to their votes. In cases where voters do not have digital certificates, a key roaming mechanism can be used to provide them with a certificate, protected by a voter-known PIN or password, which is not stored in a remote database and cannot be used to impersonate the voter.

Vote Encryption: In an electronic voting platform, the confidentiality of votes can be compromised during transmission and storage. To ensure vote privacy, vote encryption at the time of casting is essential. Some voting systems apply encryption at the network level using SSL connections between the voter's computer and the voting server. However, this method is not sufficient to protect end-to-end voter privacy because the vote is not encrypted when leaving the transmission channel and arrives at the voting server in clear text, leaving it vulnerable to any attacker who gains access to the server system. To overcome this issue, it is recommended to encrypt the votes using the election public key at the data level. This way, even if an attacker gains access to the voting server, voter privacy remains intact as the votes leave the transmission channel to remain encrypted.

Protection of the Election Private Key: To protect voters' privacy and maintain the secrecy of intermediate results, asymmetric encryption algorithms are often used in the voting process. These algorithms allow for the votes to be encrypted with a public key, and only decrypted with the corresponding private key. To ensure one individual cannot misuse the private key, it should be split into multiple shares using threshold cryptography with each share given to a different electoral board member. This ensures that a minimum number of board members must work together to recover the private key and decrypt the votes. The use of a threshold scheme is crucial to prevent the loss of one share from causing the failure to decrypt the votes.

Anonymizing Votes before Decryption: Mixnets are a type of cryptographic protocol used to ensure the privacy and integrity of digital communication. In the context of an election, Mixnets can be used to protect the confidentiality of votes while still allowing for the accurate tabulation of results. The protocol works by shuffling and encrypting the votes multiple times before they are decrypted and revealed. This makes it nearly impossible for anyone, including the election administrators, to link a specific vote to a specific voter. Another approach to maintaining the privacy and integrity of elections is through homomorphic tallying. This method involves calculating election results without decrypting individual votes. Instead, each vote is encrypted and added together in a way that preserves the privacy of the individual votes. This allows for a secure and accurate tally of election results while still protecting the confidentiality of the voters. Both Mixnets and homomorphic tallying are important tools in ensuring fair and transparent elections.

Vote Integrity: Vote security can be compromised if attackers gain access to the voting system and manipulate cast votes. One solution to prevent vote tampering is to apply a digital signature to the encrypted vote. Another solution is to use a cryptographic MAC function, but this method has security risks because the key used for the MAC function must also be known by the voting server, which opens up the possibility of an attacker creating valid integrity proofs for modified votes. To address this issue, digital signatures from voters can be used to verify both the integrity and identity of the vote. Additionally, advanced cryptography techniques such as zero-knowledge proofs can be used to confirm that the encrypted vote was actually cast by the voter. To ensure the preservation of these digital signatures and zero-knowledge proofs, they can be stored along with the votes in the digital ballot box until the time of decryption.

Individual and Universal Verification Methods: To address the concern of ensuring the accuracy of remote voting, two methods of verification can be employed. One method is *Cast as Intended Verification*, which confirms that the vote received by the voting server matches the options selected by the voter. This can be done by calculating unique codes (known as Return Codes) based on the encrypted vote and returning them to the voter for comparison. As these codes are calculated using a secret key only known by the voting server, any attempts by an attacker to deliver fake codes would be detected. The other method is "Counted as Cast Verification," which ensures that the voter's cast vote is included in the final tally. This can be achieved by providing the voter with a receipt that contains a random identifier. The voter can then verify that their vote has been counted by checking if the identifier is present in the encrypted and tallied votes. It is crucial that these random identifiers cannot be linked to clear text votes, as this would increase the risk of vote buying or coercion.

Traceability and Auditability: An effective internet voting platform must have robust traceability measures in place to ensure the reliability and accuracy of election results. All sensitive operations performed in the platform's modules should be logged and protected against tampering so that any attacks or malfunctions can be quickly detected and addressed. The logs should not include information that could compromise voter privacy. To prevent attackers from altering the logs, they should be cryptographically secured. Additionally, critical processes such as vote decryption should provide cryptographic evidence of correct performance allowing auditors to verify the election results.

Case-based Adaptation: The adoption of technology, especially in the realm of e-voting, should not be a one-size-fits-all approach. Estonia is a prime example of a country that has successfully implemented e-voting, while other developed nations continue to grapple with finding the best solution. However, it is essential to recognize that the financial investment and resources required to implement such systems are not equally feasible for all nations, especially those in underdeveloped regions of Asia and Africa. Additionally, the high-tech blockchain processing transition and consensus mechanism required for secure e-voting often consume high amounts of electricity, further adding to the cost and environmental impact of implementing such systems. Therefore, it is crucial to consider the unique circumstances and resources available to each nation when exploring technological solutions for e-voting.

Training and Awareness: When introducing new technologies to the public, it is essential to first conduct pilot projects to measure the ease of use, adaptability, and user experiences. However, in countries where education rates, especially technical education, are low, authorities must find easy ways to ensure that the majority of voters can understand and use the system without difficulty. This can be achieved through training and creating awareness campaigns that help people understand the importance and effectiveness of the new method. People are often resistant to change until they are convinced of its benefits. Therefore, it is crucial to tailor these efforts based on the country's nature, culture, and demographics. By doing so, authorities can ensure that all citizens have equal access to the electoral process and that the new technology is accessible and understandable to everyone, regardless of their educational background or technical expertise.

Open-source Platform and Biometric Identity Check: The first-ever blockchain-based e-voting system in the United States utilized existing open-source blockchain platforms, such as *Ethereum*. Ethereum is a decentralized, open-source blockchain that incorporates smart contract functionality. Smart contracts ensure the proper execution of a transparent and secure voting mechanism, thereby maintaining the integrity of the ballots. Authentication and identification of the real voter also play a significant role in this voting system. Users and participants are required to authenticate their identity, for instance, a voter using a mobile device to cast their vote can use facial recognition or fingerprint scanning for identity verification. Countries around the globe can take advantage of the available resources in the market to experiment with the safest form of e-voting. They can develop their voting platform based on their specifications, using open-source blockchain platforms as a starting point. The use of smart contracts and authentication methods ensures the transparency and security of the voting process, which is essential for a successful e-voting system [12].

REFERENCES

1 Alvarez, R. M., Levin, I., & Li, Y. (2018). Fraud, convenience, and e-voting: How voting experience shapes opinions about voting technology. *Journal of Information Technology & Politics*, 15(2), 94–105. Retrieved October 27, 2022, from https://doi.org/10.1080/19331681.2018.1460288

2 Eggers, A. C., Garro, H., & Grimmer, J. (2021). No evidence for systematic voter fraud: A guide to statistical claims about the 2020 election. *CrimRxiv*. Retrieved October 27, 2022, from https://doi.org/10.21428/cb6ab371.98f5d828

3 Skovoroda, R., & Lankina, T. (2016). Fabricating votes for Putin: New tests of fraud and electoral manipulations from Russia. *Post-Soviet Affairs*, *33*(2), 100–123. Retrieved October 27, 2022, from https://doi.org/10.1080/1060586x.2016.1207988

4 Abbasi, A. N. (2022, January). *Iran vs US: A comparative analysis of the post-election scenario.* Retrieved October 27, 2022, from http://irs.org.pk/Focus/01FocusJan22.pdf

5 Freedom House. (2022). *Freedom in the world 2022: The global expansion of authoritarian rule.* Freedom in the World 2022: The Global Expansion of Authoritarian Rule | AFSC Under the Mask. Retrieved March 25, 2023, from https://underthemask.afsc.org/update/freedom-world-2022-global-expansion-authoritarian-rule

6 Luke, T. W. (2021, February 21). *Democracy under threat after 2020 national elections in the USA: 'stop the steal' or 'give more to the grifter-in-chief?'.* Taylor & Francis. Retrieved March 25, 2023, from https://www.tandfonline.com/doi/full/10.1080/00131857.2021.1889327

7 Rejolut. (n.d.). *Top Blockchain Use Cases for Enterprises.* Rejolut.com. Retrieved December 4, 2022, from https://rejolut.com/wp-content/uploads/2022/11/Blockchain-Use-Cases.pdf

8 A. H. Trechsel, U. Gasser, F. Silva, and V. Kucherenko, "Potential and challenges of e-voting in the European Union: Think tank: European parliament," Think Tank | European Parliament. Retrieved October 27, 2022, from https://www.europarl.europa.eu/thinktank/en/document/IPOL_STU(2016)556948

9 Desouza, K. C., & Somvanshi, K. K. (2022, March 9). *How blockchain could improve election transparency.* Brookings. Retrieved November 1, 2022, from https://www.brookings.edu/blog/techtank/2018/05/30/how-blockchain-could-improve-election-transparency/

10 Buckland, R., & Wen, R. (2012). The future of e-voting in Australia. *IEEE Security & Privacy*, *10*(5), 25–32. https://doi.org/10.1109/msp.2012.59

11 Yi, H. (2019). Securing e-voting based on blockchain in P2P Network. *EURASIP Journal on Wireless Communications and Networking*, *2019*(1). Retrieved November 1, 2022, from https://doi.org/10.1186/s13638-019-1473-6

12 Puiggalí, J., Chóliz, J., & Guasch, S. (n.d.). *UOCAVA Workshop Position Paper-Best Practices in Internet Voting.* Retrieved February 8, 2023, from https://www.researchgate.net/profile/Jordi-Puiggali-2

13 Khoury, D., Kfoury, E. F., Kassem, A., & Harb, H. (2018). Decentralized voting platform based on Ethereum Blockchain. *2018 IEEE International Multidisciplinary Conference on Engineering Technology (IMCET)*. Retrieved February 9, 2023, from https://doi.org/10.1109/imcet.2018.8603050

Chapter 9

Humanitarian aid and disaster relief using blockchain

Ami Padia and Priyank Mehta

BACKGROUND

Humanitarian crises encompass a range of events that negatively impact human health, well-being, and safety. These crises have been on the rise, causing distress within communities due to armed conflicts, terrorism, natural disasters, and human-caused calamities. Over the past decade, the COVID-19 pandemic has emerged as one of the most devastating disasters resulting from human action. It has affected people worldwide, leading to widespread confinement and imposing significant economic and health burdens on the global population. The pandemic, alongside natural disasters like floods, storms, earthquakes, and human-made catastrophes such as terrorist attacks and violence, has left a traumatized world population. To assist those in need, humanitarian aid organizations provide essential provisions like food, water, shelter, medical care, and support for community recovery and rebuilding [1].

In recent times, effectively and efficiently delivering aid to those in need has become increasingly challenging for humanitarian organizations, especially as the world still grapples with the aftermath of the COVID-19 pandemic. Obstacles like limited human resources, high transportation costs, international border issues, and more have compounded the difficulties [2]. Additionally, concerns have arisen regarding the transparency and accountability of organizations toward their donors. Geopolitical factors heavily influence the amount of funding countries allocate to humanitarian aid worldwide. These factors have put significant strain on major humanitarian aid organizations.

Blockchain technology has emerged as a valuable tool in facilitating global transactions across various sectors, including banking, finance, government, healthcare, insurance, media and entertainment, automotive, retail, consumer goods, and telecommunications. It has brought transparency and efficiency to customer service in these industries. The implementation of blockchain has also helped combat corruption through its trustless and decentralized ledger system, which stores transactions across multiple nodes rather than a single node, thereby eliminating the need for trust in a single entity. Transactions are recorded within a peer-to-peer network. In the field of humanitarian aid systems, blockchain has proven exceptionally beneficial. Not only does it enable non-governmental organizations (NGOs) to provide aid to the intended recipients at reduced transaction costs, but it also allows for comprehensive tracking of aid throughout the entire process. This chapter explores the advantages and challenges of implementing blockchain in the humanitarian aid sector. It delves into the use of blockchain in disaster relief, providing a case study of humanitarian aid organizations and the effects of employing blockchain technology [3].

DOI: 10.1201/9781003462033-11

DISCUSSION

International humanitarian aid systems

In recent times, humanitarian aid systems have collaborated with governments, NGOs, and international organizations such as the United Nations to address pressing global challenges. Ongoing conflicts in countries like Syria, Afghanistan, and Ukraine have resulted in a significant increase in the number of refugees seeking assistance. However, while the demand for humanitarian aid has risen, the funding received by these organizations has declined. Over the past decade, the funding requirements for humanitarian organizations have steadily increased, with approximately 55% of these requirements remaining unmet [4]. Given the detrimental effects of humanitarian crises, it is crucial to establish effective relief mechanisms and systems. Key aspects include highly efficient and resilient supply chains, timely and effective aid distribution, transparent communication between donors and recipients, and swift identification of victims [5]. Unfortunately, the reduced funding has impacted the focus of humanitarian organizations on the entire disaster lifecycle, including mitigation, preparedness, response, and rehabilitation. Former United Nations Secretary-General Ban Ki-Moon has expressed concern that the world is currently witnessing the highest levels of human suffering since World War II [4].

A study conducted by *Bricout* and *Vincent* in 2020 revealed that as of 2017, ten countries received approximately 63% of humanitarian aid allocations, with around 30% of the funding allocated to countries like Syria, Yemen, and South Sudan. European and North American countries have been the largest contributors to these funds. Collaborative efforts among organizations are essential to coordinate aid delivery effectively and efficiently. According to the *Global Humanitarian Assistance Report*, international humanitarian aid experienced an estimated increase of 30% from 2014 to 2018, rising from US$22.2 billion to US$28.9 billion. The primary objective of all international organizations involved in humanitarian aid is to alleviate suffering, save lives, and uphold the dignity and rights of those affected by crises [6].

In the current funding system for humanitarian aid, governments of different countries serve as the primary contributors. Government organizations have two options for participating in humanitarian aid: bilateral and multilateral channels. In a bilateral channel, the donor country maintains a direct relationship with the recipient country, providing funding directly from its official government sources. Conversely, in a multilateral arrangement, an indirect relationship is established between the donor country and the recipient country. In this case, the donor country provides funds to international financial institutions such as the *World Bank, African Development Bank, Asian Development Bank, European Bank for Development and Reconstruction, Caribbean Development Bank, and Inter-American Development Bank*. These international financial institutions then reallocate the funds to the recipient countries. It is important to note that governments are not the sole providers of funds. In addition to governments, numerous private organizations, individuals, and foundations contribute significant funding to international financial organizations, often channeled through reputable NGOs. This implies that funding goes through multiple intermediaries before reaching the intended recipients [6].

Humanitarian-based blockchain systems

Utilizing blockchain technology in the international humanitarian aid sector presents several challenges and considerations. The existing literature highlights three major issues that need to be addressed when assessing the applicability of distributed ledger technologies like blockchain.

First, there is a lack of clarity regarding what constitutes "humanitarian applications" of blockchain. Second, there is insufficient evidence on reusing and building upon existing technology. Lastly, there is a knowledge gap concerning the ethics and governance of distributed ledger technologies [7].

Blockchain technology extends beyond its applications in cryptocurrency. It holds the potential to facilitate development and growth in various fields, including international humanitarian aid. By leveraging blockchain, the sector can become more efficient, transparent, and reliable. Specifically, blockchain can be employed to automate and enhance logistical operations and information infrastructures. The evolution of Blockchain 2.0 and the latest version, Blockchain 3.0, has addressed many criticisms of Blockchain 1.0, such as the increase in criminal activities within Bitcoin transactions leading to fraudulent transactions and significant losses. By incorporating moral principles and ethics into the distributed ledger code, users can align their earnings with their own values. However, concerns may arise regarding the potential misuse of blockchain for political agendas or targeted communities of recipients. For instance, donations could be digitally earmarked to reach only specific communities. It is important to note that such proceedings are not inherent to blockchain technology itself but rather stem from the decisions made by users. The advanced features of blockchain, such as trustless and decentralized ledgers that record all transactions within a peer-to-peer network, can help mitigate corruption schemes and ensure a high level of transparency in transactions. In recent times, numerous humanitarian aid and development organizations have been exploring the combination of government and private funding with blockchain technology to enhance transparency, improve efficiency, and deliver funds to those in greatest need.

Overall, blockchain technology holds promise for revolutionizing international humanitarian aid by addressing challenges related to transparency, efficiency, and corruption. However, it is imperative to navigate ethical considerations and develop appropriate governance frameworks to maximize the benefits of blockchain technology in the sector [3].

Benefits of blockchain in humanitarian operations

Blockchain technology offers several benefits in the context of humanitarian operations, addressing critical challenges and improving efficiency and accountability. The following are key advantages of using blockchain in the humanitarian aid sector.

Traceability and Transparency: Blockchain enables the traceability of funds throughout the entire supply chain. It records every transaction and action taken by intermediaries, allowing for full transparency. This traceability ensures that resources reach the intended recipients and provides unprecedented levels of transparency for reporting, auditing, and compliance [6] [8].

Increased Coordination: Blockchain revolutionizes coordination among stakeholders in the humanitarian sector. By leveraging blockchain technology, multiple donors, governments, and aid agencies can collaborate effectively. It helps avoid administrative overlap, duplication of efforts, and facilitates policy harmonization, resulting in less resource waste and more efficient interventions.

Reduced Costs: Implementing blockchain in humanitarian aid systems can significantly reduce transaction costs. By eliminating the need for multiple financial intermediaries and harmonizing coordination and procedures, costs associated with cash donations and cross-border money transfers can be lowered. For instance, the *United Nations World Food Programme* achieved a 98% reduction in money transfer costs through their blockchain pilot project, potentially saving millions of dollars [6].

Diversification of Funding Sources: Blockchain and cryptocurrencies provide new avenues for fundraising in the humanitarian aid sector. NGOs have started accepting donations in cryptocurrencies, and blockchain can serve as a crowdfunding tool for cash-based donations. Blockchain technology enables transparent tracking of donations from donors to recipients, and smart contracts ensure funds are disbursed only when specific milestones are met. Donors can audit their donations, increasing accountability [6, 9].

Fraud Reduction: The tamper-proof nature of blockchain data reduces fraud and safeguards financial aid from corrupt governments. Blockchain technology provides assurance to donors and aid organizations by ensuring the integrity of cash donations. It offers verifiability, security, and transparency, along with features such as disintermediation, smart contracts, cost and time reduction, monitoring and tracking impact, and combating fraud and corruption [6].

In short, blockchain technology offers tangible benefits in humanitarian operations by enhancing traceability, transparency, coordination, cost-efficiency, diversification of funding sources, and reducing fraud. Leveraging blockchain can improve the overall effectiveness and accountability of humanitarian aid efforts.

Challenges of implementing blockchain in humanitarian operations

Despite its many potential benefits, caution must also be exercised in objectively assessing some potential challenges of implementing blockchain technology for humanitarian and disaster relief purposes, as follows.

Lack of Willingness: The competitive nature of the humanitarian aid sector can make it challenging to attract stakeholders to adopt blockchain technology. The decision to implement blockchain-based systems and initiatives depends on the willingness and cooperation of stakeholders. Previous initiatives like the *International Aid Transparency Initiative* have faced challenges in achieving practical effects and garnering support [10].

Data Privacy and Security: With the increasing number of stakeholders involved in blockchain-based humanitarian aid systems, ensuring proper information disclosure and obtaining consent from all parties becomes crucial. Safeguarding the privacy and security of stakeholders' data is a significant challenge, considering the potential threats posed by hackers and the need to maintain confidentiality [2].

Technological Limitations: Blockchain technology relies on internet connectivity, which can be problematic in disaster-affected areas or regions with limited access to the internet. Lack of connectivity and energy consumption pose challenges for updating and utilizing blockchain technology in humanitarian operations [10].

Scalability Issues: While blockchain has the capacity to handle large volumes of transactions, scaling the technology to meet the demands of the humanitarian aid sector can take time. Adequate infrastructure, such as sufficient machines to store and process transactions, is necessary for seamless scalability [11].

Immutability Issues: The immutability of blockchain transactions can be a challenge when errors occur. While it is possible to add reversing or adjusting transactions to correct mistakes, it requires careful review and expertise before deploying them on the blockchain [11].

Structural Reluctance: Blockchain technology combines various existing technologies, resulting in a complex system structure. Setting up systems that require internet connectivity and electricity access in crisis areas can be challenging, potentially hindering the adoption of blockchain in humanitarian operations [12].

Regulatory Challenges: Regulatory compliance is crucial when designing blockchain-based systems, taking into account international humanitarian law, human rights law, and country-specific regulations. Legal considerations, such as the legality of cryptocurrencies in certain jurisdictions, can pose challenges to the widespread adoption of blockchain in humanitarian aid [6].

These challenges highlight the complexity and considerations involved in implementing blockchain technology in the humanitarian aid sector. Overcoming these obstacles will require collaboration, stakeholder engagement, technological advancements, and adherence to legal and ethical frameworks.

Disaster relief using blockchain

Blockchain technology has the potential to significantly improve disaster relief efforts in several ways.

Coordination and Supply Chain Management: Blockchain can facilitate better coordination among various organizations involved in disaster relief, such as NGOs, government agencies, and volunteers. It enables the registration and verification of volunteers, tracking and monitoring the delivery of goods and services, and ensuring transparency in the supply chain. This can help streamline the distribution of vital resources like vaccines, food, and medical supplies [13].

Trust and Transparency: Blockchain's immutability and transparency features can help restore trust in disaster-affected areas. By recording and maintaining information on the blockchain, such as the identities of repair crews, service quality statistics, and response times, it becomes easier to verify and audit the actions taken during relief operations. This can prevent fraud and provide a reliable source of information for decision-making [14].

Information Sharing and Collaboration: Blockchain provides a secure platform for sharing and accessing crucial data across different information systems and geographical boundaries. This can improve collaboration and information exchange between organizations involved in disaster response, enabling faster and more effective decision-making [13].

Accountability and Dispute Resolution: In the event of disputes or fraudulent activities, blockchain can provide an indisputable record of transactions and actions. This can expedite the resolution process by providing reliable evidence for legal, insurance, and law enforcement purposes. It also ensures that information remains intact over time, eliminating the possibility of tampering or manipulation [14].

By leveraging the capabilities of blockchain technology, disaster relief efforts can become more efficient, transparent, and accountable. However, it is important to address challenges such as infrastructure limitations, data privacy, and regulatory compliance to fully harness the potential of blockchain in disaster response.

Blockchain technology has shown great potential in assisting vulnerable demographics, particularly international refugees in several ways.

Digital Identity: Refugees often face challenges in obtaining and retaining their personal identification documents. Blockchain can provide a solution by securely storing their digital identities, allowing them to access services such as healthcare, employment, and education. By using a decentralized and tamper-proof database, refugees can have control over their identity information and regain their documentation digitally [15].

Job Opportunities: Blockchain can help refugees find employment opportunities by verifying their skills and qualifications. By creating a digital record of their education, work experience, and skills on the blockchain, refugees can provide verifiable information to potential employers, enhancing their chances of finding suitable jobs [16].

Transparent Financial Assistance: NGOs and government agencies often provide financial assistance to support refugees. Blockchain technology can ensure the transparent and accountable allocation of funds. By recording transactions on the blockchain, organizations can track the flow of funds, prevent corruption, and ensure that resources reach those in need. Cryptocurrencies and blockchain-based vouchers can be used to distribute aid in a secure and fraud-resistant manner [16].

Financial Inclusion: Initiatives like the one in Finland, where prepaid Mastercard cards were given to unbanked refugees, demonstrate how blockchain can facilitate financial inclusion. By providing a unique digital identity stored on the blockchain, refugees can access banking services, receive direct deposits, and build credit scores. This enables them to participate in the digital economy and receive benefits from the government [17, 18].

Skill Development and Integration: Blockchain can support the integration of refugees into their host countries by providing a transparent platform for sharing information between governments, citizens, and businesses. This can help identify the needs of refugees, provide appropriate solutions, and support their skill development and language learning, increasing their chances of finding employment [16].

By leveraging the benefits of blockchain technology, the unique challenges faced by international refugees can be addressed, promoting their inclusion, empowerment, and access to essential services and opportunities.

Case study 1: Building Blocks

Building Blocks is a significant implementation of blockchain technology by *the World Food Programme* (WFP) that aims to provide humanitarian assistance to millions of people in Jordan, Bangladesh, and other regions affected by conflict, disasters, and climate change. It serves as a secure and efficient platform for accessing assistance from multiple NGOs and aid organizations. Below are some key aspects of Building Blocks.

Secure Access to Multiple Forms of Assistance: Building Blocks allows individuals to access various forms of assistance, such as food, healthcare products, and financial aid, from multiple organizations through a single access point. This simplifies the process of obtaining humanitarian aid and reduces complexity for beneficiaries.

Anonymous Identity and Privacy: Personal information, including names, dates of birth, and biometrics, is not stored on the Building Blocks platform. Instead, the system uses anonymous identities to ensure the safety, security, and privacy of individuals receiving assistance. This protects the sensitive information of beneficiaries.

Fund Transfers and Financial Support: Building Blocks facilitate the transfer of funds to individuals in need. By leveraging blockchain technology, the platform ensures transparent and efficient transactions, reducing the risk of fraud and ensuring that financial support reaches the intended beneficiaries.

Pilot Programs and Expansion: Building Blocks started with successful pilot programs, such as the 100-person pilot in Sindh Province, Pakistan, where blockchain technology was used to transfer money to beneficiaries. These pilot programs have paved the way for larger-scale implementations in areas like Bangladesh, where thousands of households receive essential goods through local stores using QR codes [19].

The primary goal of Building Blocks and the World Food Programme is to provide humanitarian assistance to vulnerable populations and contribute to the achievement of zero hunger. By leveraging blockchain technology, WFP aims to streamline aid delivery, increase efficiency, and ensure that assistance reaches those who need it most. As such, Building Blocks demonstrates the potential of blockchain technology to transform the delivery of humanitarian assistance, providing a secure and efficient platform for accessing multiple forms of aid from various organizations [20, 21].

Case study 2: Rahat

Rahat is a Cash and Voucher Assistance (CVA) management system that utilizes blockchain technology to facilitate efficient and transparent humanitarian aid distribution. It aims to provide quick relief with minimal overhead costs. The main stakeholders involved in the Rahat platform are aid agencies, beneficiaries, merchant vendors, and financial institutions. Here are some key features of the Rahat application [22, 23].

Rahat Aid Agency Dashboard: This application is used by aid agencies to manage and distribute humanitarian aid. It allows agencies to create and manage beneficiary profiles, allocate funds, and track the distribution process. By leveraging blockchain technology, the dashboard ensures transparency and accountability in aid delivery.

Rahat Vendor App: The Rahat Vendor app is designed for merchant vendors who accept vouchers provided by the aid agencies. Vendors can use the app to verify vouchers, record transactions, and receive payment for the goods or services they provide to beneficiaries. The app helps streamline the distribution process and ensures that aid is used appropriately.

Rahat Social Mobilizer: The Rahat Social Mobilizer application aims to engage local communities and individuals in the aid distribution process. Social mobilizers act as intermediaries between aid agencies, beneficiaries, and vendors, helping to ensure effective communication and coordination. They can use the app to register beneficiaries, verify their identities, and facilitate the distribution of aid.

By using blockchain technology, Rahat provides several benefits to the humanitarian aid sector, such as:

Efficiency and Transparency: Blockchain technology ensures efficient and transparent aid distribution by recording all transactions on a decentralized ledger. This helps prevent fraud, corruption, and duplication of efforts.

Cost Reduction: By eliminating the need for intermediaries and reducing administrative overhead, Rahat minimizes costs associated with aid distribution. This allows a larger portion of the funds to be directed toward assisting beneficiaries.

Financial Inclusion: Rahat addresses the challenge of providing financial aid to individuals without bank accounts. By utilizing blockchain-based vouchers, beneficiaries can access aid and make transactions with participating merchants, even without a traditional bank account [23–25].

Open-Source Approach: Rahat is an open-source platform, which means it can be accessed and improved upon by a wider community of developers and organizations.

This promotes collaboration, innovation, and customization to meet the specific needs of different regions and contexts [25].

The Rahat Aid Agency Dashboard is a web application that allows aid agencies to manage and monitor the aid distribution process. With the Rumsan Wallet Application (RWA), aid agencies can log in to their account and initiate aid distribution by creating tokens from the wallet. The dashboard enables aid agencies to perform various tasks such as project creation, budget allocation, beneficiary registration, token creation and burn, vendor onboarding and management, real-time fund flow monitoring, financial service provider integration, and project reporting.

The *Vendor Application* is used by vendors and merchants participating in the Rahat aid distribution process. Vendors receive tokens from beneficiaries in exchange for the goods or services they provide. They can redeem the received tokens for cash from participating financial service providers or aid agencies. The Vendor Application allows vendors to receive tokens, redeem tokens, view transactions, transfer tokens, and check their total balance [26].

The Social Mobilizer Application is used by field volunteers of the Rahat Foundation to register beneficiaries and provide them with tokens on behalf of the aid agencies. The tokens can be used by beneficiaries to purchase commodities from participating vendors or exchange them for cash from the aid agencies. The Social Mobilizer Application enables social mobilizers to register beneficiaries, view the list of registered beneficiaries, assign tokens, and check the budget balance of the current project.

Each aid agency is provided with a digital wallet, which acts as a unique identifier for logging into the Rahat application. The Rahat ecosystem simplifies and tracks the aid distribution process. Aid agencies can perform blockchain-based transactions using the *Rumsan* wallet, including registering recipients, creating projects, onboarding vendors, and allocating relief funds. Recipients receive relief tokens via text messages, which they can use to connect with Rahat vendors and receive cash or purchase items as part of the humanitarian aid.

As such, Rahat leverages blockchain technology to create an efficient, transparent, and inclusive system for humanitarian aid distribution, benefiting aid agencies, beneficiaries, vendors, and financial institutions involved in the process.

Compared to traditional cash assistance processes that were labor-intensive, slow, expensive, opaque, and prone to accountability issues, the Rahat application offers increased visibility, efficiency, and manageability. It reduces the need for manual labor, improves distribution rates, and provides better transparency. Additionally, being an open-source platform, Rahat has a greater social impact. The platform follows a Continuous Integration and Continuous Development (CI/CD) approach for mass deployments, and a public communication channel is created on the Discord platform to enhance transparency in communication [26].

The entry of *UNICEF* as an investor in the Rahat application has significantly increased its credibility. This partnership has generated global interest in the platform, and Rahat has been able to collaborate with other major organizations such as the *Ethereum Foundation*, *Inter-American Development Bank (IDB)*, *Cash Learning Partnership (CALP)*, *Nepal Red Cross Society (NRCS)*, and *Cash Coordination Group Nepal (CCGN)*.

Rahat has initiated pilot programs across Nepal to provide underprivileged communities with opportunities for growth and global visibility. Through these programs, Rahat is also promoting digital literacy, making people aware of the positive impact technology can have on the humanitarian aid distribution process. The system architecture of Rahat is robust and well-built. It consists of ten components: Rahat API, Rahat agency app, Rahat mobilizer app, Rahat vendor app, Rahat wallet, Rumsan wallet app, Rahat contracts, Rahat OTP, Rahat aid connect, and Rahat fundraise. These components work together to enable the smooth functioning of the aid distribution system.

Rahat utilizes ERC-20 tokens, which are issued on the Ethereum blockchain network. Ethereum is a decentralized platform that provides immutability, irreversibility, transparency, low cost, and decentralized security. ERC-20 tokens follow certain rules and technical specifications defined in the ERC-20 protocol standard, allowing them to be transferred within the Ethereum network and used in multiple decentralized applications. The token transactions and verification are managed on an EVM-based blockchain network. Each token is linked to a blockchain address and has its own set of rules encoded in a smart contract. Aid agencies create a list of beneficiaries, and each beneficiary has a unique account linked to the blockchain, enabling the tracking of their activities.

After the implementing partner registers the beneficiary, the beneficiary receives a one-time password (OTP) along with their account details. With the tokens received in their account, beneficiaries can collect cash or commodities from vendors in exchange for tokens. The transaction is finalized when the vendor enters the transaction details, including the OTP provided by the beneficiary. The token is then destroyed by the aid agency to prevent replication and fraud. To redeem the tokens for cash, the vendor goes to the respective bank and provides the OTP. The bank verifies the OTP and completes the transaction, providing the local currency in return [27].

Overall, the Rahat system architecture ensures secure and transparent aid distribution using blockchain technology, providing traceability and efficiency in the process.

A proposed architecture for humanitarian use cases

The proposed architecture for humanitarian use cases leverages Ethereum as the underlying blockchain technology. It incorporates various key components to enhance efficiency, transparency, and accountability in humanitarian aid delivery. Here is an overview of the proposed architecture:

Decentralized Database: Ethereum's decentralized database allows for real-time tracking, monitoring, and sharing of aid supplies, donations, and transactions. This eliminates data silos, enhances data privacy and security, and improves transparency and accountability [28].

Smart Contracts: Smart contracts are self-executing programs that automate processes and enforce predefined rules. They can be used to automate the delivery and distribution of aid supplies, disburse funds, and verify compliance. Smart contracts reduce the need for intermediaries, minimize transaction costs, and expedite aid delivery [29].

Digital Identity: Blockchain-based digital identity systems provide unique identifiers for beneficiaries, aid workers, and other stakeholders. These systems enhance identity authentication, reduce fraud, improve access to aid, and protect personal data [30].

Cryptocurrencies: Ethereum's native currency can facilitate fast, secure, and low-cost transactions for aid donations and disbursements. Blockchain-based cryptocurrency systems minimize the risks of corruption, fraud, and money laundering, ensuring aid reaches its intended beneficiaries.

Supply Chain Management: The architecture utilizes the Ethereum blockchain to store supply chain data. This enables tracking and monitoring of aid supplies from origin to delivery, improving supply chain visibility, reducing waste, and ensuring efficient distribution to those in need.

Distributed Networks: The architecture employs distributed networks to facilitate real-time collaboration and coordination among aid organizations, government agencies, and stakeholders. This enhances information sharing, situational awareness, and enables effective humanitarian responses [31].

By leveraging Ethereum's capabilities, the proposed architecture aims to streamline humanitarian aid processes, eliminate inefficiencies, and enhance the impact of aid delivery.

CONCLUSIONS AND RECOMMENDATIONS

The integration of blockchain technology in the humanitarian aid sector has shown great potential in addressing challenges related to transparency, accountability, and efficiency. By leveraging blockchain's features such as decentralized databases, smart contracts, digital identity, cryptocurrencies, supply chain management, and distributed networks, organizations can enhance the delivery of humanitarian aid to those in need.

Blockchain technology offers benefits such as increased traceability and transparency, improved coordination among stakeholders, reduced operational costs, diversification of funding sources, and fraud reduction. However, there are challenges that need to be addressed, including data privacy and security concerns, technological limitations, scalability issues, immutability challenges, structural reluctance, and regulatory considerations.

The proposed architecture based on Ethereum blockchain technology provides a decentralized, reliable, transparent, and scalable solution for delivering humanitarian aid. Ethereum's public blockchain nature and its powerful smart contracts make it suitable for a wide range of applications. Additionally, the strong support and community around Ethereum contribute to its growth and development.

By embracing blockchain technology and leveraging the proposed architecture, humanitarian aid organizations can establish efficient systems that utilize the potential of technology to provide fast, reliable, transparent, and scalable aid services. This alignment of technology and human resources can contribute to tackling challenges like hunger and poverty resulting from humanitarian crises on a global scale.

To ensure successful implementation, it is important to address concerns such as lack of clarity, knowledge gaps, data privacy and security, technological limitations, and regulatory issues. Collaborative efforts between humanitarian organizations, governments, technology experts, and stakeholders are necessary to overcome these challenges and fully realize the potential of blockchain in the humanitarian aid sector.

Suggested good practices

As a final recap, the following section re-emphasizes the five best practices for utilizing blockchain technology, especially in the context of humanitarian aid. These practices aim to optimize the data structure of blockchain systems and enhance their effectiveness in facilitating aid distribution.

> *Decentralized Database*: To establish trustless and transparent transactions, it is crucial to implement a decentralized database. This ensures that data is stored and verified across multiple nodes in the blockchain network, reducing the reliance on a centralized authority. By distributing the database, transparency is increased, as each participant can access and verify transaction records independently [1].
>
> *Smart Contracts*: Smart contracts play a pivotal role in automating and streamlining the distribution of humanitarian aid. These self-executing programs operate on the blockchain and facilitate the enforcement and fulfillment of predefined rules and conditions. By utilizing smart contracts, the need for intermediaries is minimized, leading to increased efficiency and reduced costs [2]. For instance, smart contracts can be programmed to trigger the release of aid funds or supplies based on specific criteria being met.

Digital Identity Solutions: Incorporating digital identity solutions ensures the authenticity and accountability of all participants involved in the aid distribution process. Blockchain-based digital identity systems enable individuals to have a unique identifier that can be authenticated and verified, reducing the risk of identity fraud. This not only enhances the efficiency of aid distribution but also safeguards personal data and privacy [3].

Cryptocurrency Integration: The use of cryptocurrencies within blockchain systems offers several advantages for humanitarian aid. By leveraging blockchain-based cryptocurrencies, such as Ethereum, organizations can diversify their funding sources and reduce transaction costs. Furthermore, the transparent and traceable nature of cryptocurrency transactions enhances accountability and ensures that aid reaches its intended beneficiaries [4]. Implementing token standards like ERC-20, ERC-721, ERC-777, and ERC-1155 can provide additional functionalities and compatibility within the blockchain ecosystem.

Privacy and Security Considerations: Privacy and security are paramount in humanitarian aid operations. Blockchain systems must incorporate robust data protection mechanisms to safeguard sensitive information. Additionally, organizations should adhere to ethical and regulatory frameworks governing data privacy and security. Implementing encryption techniques, permissioned access controls, and consensus mechanisms that prioritize security can help address privacy concerns and protect against unauthorized access [6].

It is important to note that scalability and immutability issues should be carefully considered when designing blockchain solutions for humanitarian aid. As the volume of transactions and data grows, the blockchain infrastructure must be capable of handling increased demands without compromising performance. Collaboration with stakeholders and community members is crucial to ensure successful adoption and implementation of blockchain technology, as it requires active participation and support from all involved parties [7].

By adhering to these best practices and leveraging the potential of blockchain technology, the humanitarian aid sector can overcome challenges related to transparency, accountability, and efficiency. The proper implementation of blockchain solutions has the potential to revolutionize aid distribution, enabling faster, more reliable, and transparent assistance to those in need.

REFERENCES

[1] A. Narayanan, K. Hunt and J. Zhuang, "Blockchain in humanitarian operations management: A review of research and practice," *Socio-Economic Planning Sciences*, vol. 80, 2022.

[2] K.R. Cross, "Blockchain Technology in Humanitarian Programming," *International Center for Humanitarian Affairs*, 2018.

[3] A. Zwitter and B.-D. Mathilde, "Blockchain for humanitarian action and development aid," *Journal of International Humanitarian Action 3*, vol. 16, 2018.

[4] M. Besiou and L. N. V. Wassenhove, "Humanitarian Operations: A World of Opportunity for Relevant and Impactful Research," *Manufacturing & Service Operations Management*, vol. 22, no. 1, pp. 135–145, 2020.

[5] M. Pisa and M. Juden, "Blockchain and Economic Development: Hype vs. Reality," *Center for Global Development*, 2017.

[6] A. Bricout and A. Vincent, "Solving Humanitarian Aid Inefficiencies with Blockchain Technology," *Frankfurt School Blockchain Center*, August 2020.

[7] G. Coppi and L. Fast, "Blockchain and Distributed Ledger Technologies in the Humanitarian Sector," *HPG Commissioned Report*, 2019.

[8] Blockchain for Development: Emerging Opportunities for Mobile, Identity and Aid, GSMA.

[9] Stanford Graduate School of Business – Center for Social Innovation, "Blockchain for Social Impact Moving Beyond the Hype," *Center for Social Innovation, RippleWorks*, pp. 60–68, 2018.

[10] T. Riani, "Blockchain For Social Impact in Aid And Development," April 2020. [Online]. Available: https://humanitarianadvisorygroup.org/blockchain-for-social-impact-in-aid-and-development/.

[11] G. Habib, S. Sharma, S. Ibrahim, I. Ahmad, S. Qureshi and M. Ishfaq, "Blockchain Technology: Benefits, Challenges, Applications, and Integration of Blockchain Technology with Cloud Computing," *MDPI*, vol. 14, no. 11, 2022.

[12] S. N. Khan, F. Loukil, C. Ghedira-Guegan, E. Benkhelifa and A. Bani-Hani, "Blockchain smart contracts: Applications, challenges, and future trends," *Peer-to-peer Networking and Applications*, vol. 14, pp. 2901–2925, 2021.

[13] N. Das, S. Basu and S. Das Bit, "ReliefChain: A blockchain leveraged post disaster relief allocation system over smartphone based DTN," *Peer-to-peer Networking and Applications*, vol. 15, no. 6, pp. 2603–2618, 2022.

[14] M. Demir, A. A. Mashatan, O. Turetken and A. Ferworn, "Utility Blockchain for Transparent Disaster Recovery," *2018 IEEE Electrical Power and Energy Conference (EPEC)*, 2018.

[15] M. J. Morrow, M. Kovarski and A. Alfawakheeri, "The Promise of Blockchain and Safe Identity Storage for Refugees," 2018. [Online]. Available: https://www.unhcr.org/blogs/wp-content/uploads/sites/48/2018/04/fs.pdf.

[16] A. S. Bayram, "Here are three ways blockchain can change refugees' lives," 25 June 2018. [Online]. Available: https://www.weforum.org/agenda/2018/06/three-ways-blockchain-change-refugees-lives/.

[17] J. Wilmoth, "Finland Turns to Blockchain to Help Unbanked Refugees," 8 September 2017. [Online]. Available: https://www.ccn.com/finns-turn-to-blockchain-to-help-unbanked-refugees-enter-the-digital-economy/.

[18] R. Juskalian, "Blockchain technology helps Syrian refugees regain lost legal identities," 2018. [Online]. Available: https://www.business-humanrights.org/en/latest-news/blockchain-technology-helps-syrian-refugees-regain-lost-legal-identities-2/.

[19] V. Matak, "How blockchain can power efforts to empower women and girls in Bangladesh," 13 December 2022. [Online]. Available: https://www.wfp.org/stories/how-wfps-blockchain-platform-powers-efforts-empower-women-and-girls-bangladesh#:~:text=WFP%20uses%20the%20blockchain%2Dbased,by%20the%20UN%20Population%20Fund.

[20] wfp, "Building Blocks | WFP Innovation," 15 February 2022. [Online]. Available: https://innovation.wfp.org/project/building-blocks.

[21] Rumsan, "Rahat: Blockchain for faster, transparent cash assistance for beneficiaries in hard-to-reach areas," 1 September 2022. [Online]. Available: https://www.unicefinnovationfund.org/broadcast/updates/rahat-blockchain-faster-transparent-cash-assistance-beneficiaries-hard-reach.

[22] Esatya, "Rahat: An open-source, blockchain based, Humanitarian Aid distribution management system," 21 June 2022. [Online]. Available: https://esatya.io/blogs/rahat-an-open-source-blockchain-based-humanitarian-aid-distribution-management-system.

[23] rahat, "Rahat - Aid Token Distribution," [Online]. Available: https://www.rahat.io/platformDetail/.

[24] P. Gyawali and S. B. Chhetri, "https://rahat.io/blogs/rahat-aid-agencies-transforming-the-humanitarian-aid-distribution-using-Blockchain-technology/," 25 August 2021. [Online]. Available: https://rahat.io/blogs/rahat-aid-agencies-transforming-the-humanitarian-aid-distribution-using-Blockchain-technology/.

[25] S. B. Chhetri, "Fostering Digital Literacy in Humanitarian Aid through Rahat," 30 April 2022. [Online]. Available: https://rahat.esatya.io/blogs/fostering-digital-literacy-in-humanitarian-aid-through-rahat/.

[26] Rahat, "System Architecture | Rahat - Aid Token Distribution," 2022. [Online]. Available: https://docs.rahat.io/docs/architecture/.

[27] Rahat, "Relief distribution using Blockchain-based tokens," [Online]. Available: https://assets.rumsan.com/esatya/rahat-whitepaper.pdf.

[28] S. Kennedy, "Exploring Decentralized Architecture Networks in Healthcare," 09 January 2023. [Online]. Available: https://healthitanalytics.com/features/exploring-decentralized-architecture-networks-in-healthcare.

[29] ISDA, "Smart Contracts and Distributed Ledger – A Legal Perspective," August 2017. [Online]. Available: https://www.isda.org/a/6EKDE/smart-contracts-and-distributed-ledger-a-legal-perspective.pdf.

[30] Cloudflare, "Ethereum Network | Cloudflare Web3 Docs," 01 August 2022. [Online]. Available: https://developers.cloudflare.com/web3/ethereum-gateway/concepts/ethereum/.

[31] Crypto, "What are Token Standards? An Overview," 02 February 2022. [Online]. Available: https://crypto.com/university/what-are-token-standards.

Chapter 10

Enhancing healthcare in underserved regions through blockchain

Lilian Behzadi and James Joseph

BACKGROUND

Telemedicine, a remote medical care service, enables patients to access healthcare services quickly and accurately, particularly in hard-to-reach regions. It facilitates the presence of experts and specialists in both remote and non-remote areas, allowing healthcare professionals to connect directly with specialists and provide necessary treatments [1]. Although telemedicine has existed for a considerable period, its progress has been hindered by financial, regulatory, and technical challenges. Legislative initiatives like the American Recovery and Reinvestment Act (ARRA) and the Health Information Technology for Economic and Clinical Health (HITECH) Act have brought about advancements in telemedicine by streamlining the exchange of patient health information and promoting the use of electronic medical records [2].

The roots of telemedicine can be traced back to the 19th century when a telecommunication infrastructure, including the telephone, radio, and telegraph, was established. During the Civil War, telegraph communication enabled the reporting of injuries and casualties and facilitated consultations with doctors and the ordering of medical supplies. These early applications represented the initial adoption of telemedicine technology. In 1879, it was demonstrated that telephone usage could reduce patient visits to clinics and hospitals. Dr. *Hugo Gernsback* envisioned in 1922 the use of sensory feedback devices that would enable medical personnel to communicate with patients through screens and perform examinations using robotic arms. By 1948, medical staff in Pennsylvania received the first radiologic images via telephone lines, and neurological examinations were transmitted across a university campus using two-way interactive television. In subsequent years, closed-circuit television links were established, allowing physicians to provide consultations over 100 miles away. Today, in the 21st century, telemedicine devices such as computers, phones, and tablets are readily accessible to most individuals. This accessibility enables patients in rural and busy areas to connect with healthcare providers easily, while homecare devices allow medical personnel to monitor various aspects of their patients' health, from vital signs to glucose levels. Telemedicine enables doctors, specialists, and other medical professionals to provide diagnoses and information to patients who are unable to physically visit clinics and hospitals [3].

Telemedicine services are facilitated through interconnected platforms that enable the exchange of data to provide healthcare services to patients in regions where medical personnel face challenges in reaching them. Medical teams can upload data and information, such as images, to these platforms which specialists can access to provide diagnoses. The "store and forward" model involves sending data to specialists who offer additional insights to doctors and other medical personnel in the field. This system ensures patient privacy and data protection, reducing the isolation experienced by international and local healthcare

DOI: 10.1201/9781003462033-12

providers. Ultimately, the emphasis on privacy contributes to higher service quality and improved access, particularly in remote areas. Telemedicine also offers easy access to experts, treatments, and specialists, with quick responses to medical inquiries typically within 24–48 hours. *Médecins Sans Frontières (MSF)*, AKA *Doctors Without Borders*, for instance, delivers medical care through a network of experts using various devices supported by a 24-hour MSF center [4].

In 2010, MSF implemented Telemedicine in its field operations aiming to overcome barriers and improve access to healthcare. This innovative service offers enhanced speed and accuracy in delivering medical care to regions that are challenging to reach. By leveraging telemedicine, medical professionals, including doctors, nurses, specialists, and other personnel can be virtually present in these areas. The telemedicine software utilized by MSF ensures efficient management of medical records and facilitates confidential consultations. This system incorporates appropriate technologies to provide services through both portable and non-portable devices, supporting the operations of MSF [1].

DISCUSSION

Integration of blockchain and telemedicine

The combination of blockchain technology with telemedicine offers several advantages addressing key challenges and enhancing the overall efficiency and security of healthcare systems. One crucial factor is centralization, a fundamental aspect of telemedicine. Blockchain's distributed architecture enables the creation of a shared ledger for health records, ensuring synchronization and verification across all nodes. Blockchain technology effectively tackles challenges such as tracking medications and drugs, verifying physician credentials, and monitoring patient locations. By addressing issues like privacy breaches and counterfeit drugs, blockchain technology contributes to the growth of the healthcare market, with the global blockchain health market projected to reach $829 million by 2023.

Blockchain technology plays a significant role in enabling the telemedicine sector by providing secure information sharing, verifiable data, and maintaining health information privacy. The COVID-19 pandemic accelerated the adoption of telemedicine as it became essential to deliver healthcare virtually and remotely to ensure social distancing. The telemedicine market, valued at $45 billion in 2019, is projected to grow to over $290 Billion in 2030. Blockchain technology facilitates the development of applications that offer convenient and consolidated views of Electronic Health Records (EHRs), ensuring transparency and visibility of patients' medical history. It enables secure information sharing, verifiable data, and privacy [5].

Another important feature of blockchain technology is its decentralization, which enhances the robustness of existing systems, safeguarding patient EHRs against malicious attacks and unintentional data loss. Consensus protocols ensure that all involved parties reach an agreement on the current state of the ledger. Every transaction in blockchain-based systems is digitally signed using public key cryptography, ensuring the immutability of medical records. Blockchain technology can cater to the specific needs of the healthcare sector, facilitating quick and real-time sharing of EHRs, efficient patient health data management, low-cost and high-performance solutions, data security, privacy, availability, and establishing the provenance of health records. Additionally, blockchain technology can safeguard interactive medical procedures by preventing database manipulation and ensuring tamper-proof logs. It enables the effective and reliable automation of telemedicine services through the use of smart contracts [6].

MSF scope of services

Examples of non-profits that provide telemedicine to undeveloped countries include *Médecins Sans Frontières (MSF)*, aka *Doctors Without Borders, UNICEF,* and *The Red Cross.* MSF uses telemedicine software that allows medical personnel to manage medical information remotely and medical consultations with maximum confidentiality. The system uses technologies to offer secure services through portable and non-portable devices [7]. MSF also offers innovative programs and tools for saving lives, like solar power to help the most vulnerable, epidemiology with a smartphone, safer cooking methods in conflict zones, three-wheeled urban ambulances, and advanced care to places where specialists can't go. Solar power can help the most vulnerable patients by using concentrated oxygen, an essential element that is important to provide care to sick patients. Concentrated oxygen is, however, a constant challenge to transport and extremely expensive to produce. Note that while air contains 20% oxygen, concentrated oxygen are filled with a 90% mixture. Oxygen concentrators are used, but a great deal of energy is required to operate the concentrators. About one hundred liters of diesel are required to power one oxygen concentrator. The oxygen concentrators use most of the energy in MSF projects and are considered primary energy users. MSF has developed an alternative cooking tool called briquettes to provide safer cooking methods. Blocks of fuel are made by pressing biowaste, which results in a final product called briquettes. Biowaste can include groundnut bark or sugarcane peel mixed with paper. Approximately about 600 houses have replaced the use of firewood with briquettes, and more individuals are trying to create the same process on their own. Globally, three-wheeled vehicles are a convenient and cheap mode of transportation. In some places, vehicles are used more than just for the transportation of goods. The three-wheelers are also used to save lives. About 260 patients are transported to hospitals every month. MSF uses the three-wheelers as temporary ambulances to avoid movement restrictions in certain places. MSF provides nutritional and inpatient care, emergency consultations, and sexual reproductive health services [8].

Remote care through other telemedicine initiatives

To enhance the proficiency of family physicians in utilizing technology for remote care, a series of training workshops is being supported by *UNICEF* and the *Macedonian Medical Association.* Telemedicine, which involves using a telecommunications network to provide care to patients at different locations, is employed by doctors to bridge the gap between them and their patients. The objective is to ensure equal access to healthcare for all families, regardless of their location and to offer alternatives to traditional face-to-face consultations when physical presence is not feasible or necessary [9].

Amid the COVID-19 pandemic, the European Union (EU) and four United Nations organizations – namely, *the World Health Organization* (WHO), *United Nations Population Fund* (UNFPA), *UNICEF,* and *United Nations Office for Project Services* (UNOPS) – joined forces with the Ministry of Internally Displaced Persons to launch a digital health initiative. The *Ministry of Internally Displaced Persons* is an entity or department responsible for addressing the needs and concerns of individuals or groups who have been forced to flee their homes due to conflicts, natural disasters, or other reasons within their own country. The objective of this initiative is to minimize the impact of the pandemic on internally displaced persons and provide them with necessary healthcare services As part of the effort, 200 rural facilities would receive essential equipment, while an additional 50 would acquire telemedicine technology. Simultaneously, routine pediatric medical services would be provided and the new equipment would ensure continuous access to healthcare for individuals with chronic diseases. By engaging healthcare providers in online training, the initiative seeks to enhance and expand the delivery of high-quality primary healthcare services. Moreover, the project aims to establish a strong and sustainable foundation for digital health and telemedicine by supporting the development of legal and regulatory frameworks, clinical guidelines, and specific training programs for healthcare personnel. Additionally, it seeks to promote the utilization of telemedicine services among healthcare providers and the general public [10].

The *UNICEF Venture Fund* highlights a Brazilian startup called *Portal Telemedicina*, which has developed a platform that offers fast, reliable, and affordable diagnostics in over 300 cities. The platform integrates AI technology to provide medical insights to healthcare providers and enables doctors to conduct online diagnoses. Portal Telemedicina serves as a telehealth and population health management platform that connects medical devices and records, transmitting data for doctors to access patient files and conduct remote consultations. AI models trained on millions of examinations are utilized to identify abnormalities and prioritize emergencies. The technology developed by Portal Telemedicina focuses on population health management in the field of child development, creating a platform tailored to the needs of children, teenagers, and pregnant and postpartum women. The platform aims to integrate various datasets, encompassing health, education, and socioeconomic aspects of child development to provide a comprehensive perspective on the health of pregnant women and children. This will enable informed decision-making by public managers regarding corrective and preventive measures. The platform includes an AI module that generates insights and alerts based on key markers of child development. The central database ensures unified and accessible data for intelligent data analysis with algorithms facilitating data cleansing, enrichment, and duplication. The scalable database design allows for future expansion and increased user capacity. The integration of databases related to different areas of child development proves to be particularly valuable. The AI layer for alerts enables in-depth analysis in conjunction with database integration [11].

Red cross and Teladoc partnership: bridging the gap in disaster healthcare

The Red Cross has joined forces with *Teladoc*, a Texas-based corporation, to establish a digital connection to healthcare for victims of disasters. Through this partnership, victims gain access to Teladoc's phone and web-based services using a dedicated app. This enables them to communicate with Teladoc doctors during and after the incident. A press release highlights that telehealth is expected to play a significant role in the future of disaster risk reduction. The collaboration between these two organizations aims to address the service gap that arises during events and continues in the subsequent hours, days, and weeks. Teladoc is just one of the many organizations leveraging mobile health technology to provide emergency and triage care. Often, this involves equipping first responders with the ability to connect with medical professionals through an app or telehealth link [12].

Advantages of blockchain technology in telemedicine

As presented in Table 10.1, Blockchain technology brings several advantages to the field of telemedicine, including patient control over personal information, streamlined patient identification, and improved collaboration among healthcare providers.

Patient Control over Information: With Blockchain, patients have greater control over their medical records and personal information. These records are securely stored on the Blockchain and shared virtually with healthcare providers as needed. Whenever there are updates to the records, all registered devices on the network receive the updated

Table 10.1 Summary of advantages and disadvantages of blockchain in healthcare [4, 13]

Entities	Advantages/Disadvantages
Healthcare organizations	Improved decision-making, allowing multiple doctors to view the same records from different locations in real time. Tamper-proof medical records as they're transformed into a decentralized system. Quick credentialization process.
Patients	Blockchain enables patients to assume ownership of their information. Consensus mechanisms prohibit medical personnel from accessing records without permission or knowledge. Patients can participate in research studies, and the collection and storage of data from different devices are done securely.
Pharmaceuticals	Recruitment of individuals for clinical trials is enabled using Blockchain technology. Reliable and auditable documentation is enabled with immutable records and counterfeit drug detection. Distributed ledgers are created to maintain transparency as medicines are transported across the country.
Insurance	Customer Confirmation process is quicker. Automated payments for claims. Automated coverage verification between companies and insurers. Lower administrative costs. Smart contracts can manage agreements between different parties.
Healthcare Organizations, Patients, Pharmaceuticals, Insurance	Increase in transaction time causing delays as the number of blocks added to the chain grows. Medical records can't be permanently deleted. New blocks are added to indicate the invalidation of existing blocks. Problems with large files, e.g., MRI scans. Higher development and maintenance cost when using Blockchain technology.

version. This empowers patients to manage their health information and ensures its accuracy and integrity.

Streamlined Patient Identification: Blockchain technology simplifies the process of identifying patients, particularly when they undergo changes in their personal information. Whether it is a change in name, phone number, address, or medical providers, the Blockchain ensures that patient data remains secure and uncompromised. By eliminating duplication, deletion, and discrepancies in patient records, Blockchain enhances the efficiency and accuracy of patient identification.

Enhanced Collaboration among Providers: Healthcare providers benefit from the centralized nature of Blockchain-based healthcare data records. This technology enables unified and coordinated information sharing among healthcare teams. As a result, providers can access comprehensive patient data, leading to more timely and accurate diagnoses. The collaborative environment fostered by Blockchain facilitates seamless communication and cooperation among healthcare professionals.

Health Supply Chain Management: Blockchain technology can be utilized to improve the efficiency, transparency, and security of healthcare supply chains. It enables tracking and tracing of pharmaceuticals, medical devices, and other healthcare products throughout the supply chain, ensuring authenticity and reducing the risk of counterfeit or substandard products. Blockchain-based systems can also streamline inventory management, automate order processing, and enhance supply chain visibility, leading to improved patient safety and quality of care. Supply Chain Management (SCM) in healthcare focuses on improving the delivery of healthcare products, medications, and resources. However, distributed ordering poses risks, including the distribution of counterfeit drugs from untrustworthy suppliers. Blockchain technology offers a monitoring tool by providing access to every stage of the transportation process through a distributed ledger. This enables verification of medication origin, supplier credibility, and distributor authenticity. Healthcare professionals can utilize the blockchain to confirm suppliers' credentials, ensuring that genuine medications reach patients in need, thereby enhancing supply chain transparency and authentication [14].

Supply Chain Transparency: Blockchain technology enables transparency in supply chains by recording and verifying every transaction and movement of goods. This transparency can help prevent fraud, ensure ethical sourcing of pharmaceuticals and medical devices, and provide greater visibility into the origin, quality, and handling of products. Patients and healthcare providers can have confidence in the authenticity and integrity of the supplies they receive. Supply chain transparency is a crucial aspect of healthcare particularly in addressing the issue of counterfeit drugs. Developing countries face a significant challenge with an estimated 1% of drugs being counterfeit, rising to over 10% in some regions. The prevalence of counterfeit drugs leads to numerous deaths each year [13]. Furthermore, as the use of medical devices and remote monitoring increases, ensuring the authenticity and integrity of these devices becomes increasingly important, as they can be targeted by attackers [15].

As mentioned previously, Blockchain technology offers a solution to enhance transparency in the healthcare supply chain. It enables patients to track medications and other products throughout the entire supply chain, including manufacturing, wholesale distribution, and shipping. By creating a ledger for each drug, the blockchain documents its point of origin and records every change that occurs during the supply chain process. This includes adding additional information such as labor costs and waste generated during transportation and manufacturing. Organizations like *MediLedger* and *Blockpharma* utilize blockchain protocols to verify information related to a medication's supply chain, including expiration dates, manufacturers, ingredients, and other relevant details. Blockpharma, for instance, has developed an application that scans the drug's supply chain at every shipment stage, allowing patients to verify the legitimacy of their purchase. Studies have shown that Blockpharma has the potential to intercept 15% of counterfeit drugs globally. By leveraging blockchain technology, healthcare stakeholders can enhance transparency, traceability, and authenticity in the supply chain, helping to mitigate the risks associated with counterfeit drugs and ensuring patient safety.

Faster Medical Credentialing: Traditional medical credentialing processes, which involve verifying the qualifications and credentials of healthcare professionals, can be time-consuming and prone to errors. Blockchain technology can streamline and automate this process by securely storing and sharing verified credentials on a decentralized ledger. This allows for faster, more efficient, and reliable verification of healthcare professionals' qualifications, enabling quicker onboarding and deployment in emergency situations or when rapid access to healthcare providers is required. Key advantages and benefits of using Blockchain for credential verification include improved efficiency, as Blockchain technology automates and streamlines the credential verification process, reducing the time and effort required. This eliminates the need for lengthy phone calls and email exchanges, resulting in faster verification timelines. In addition, the current process of maintaining healthcare provider databases can be expensive. By using Blockchain, organizations can reduce costs associated with manual verification processes, data storage, and maintenance of centralized databases. Blockchains also allow for incremental updates to credentials, making it easier for professionals to add new qualifications or update existing ones. This ensures that the credential data is always up to date and accurate. Finally, Blockchain technology enables organizations to log and monitor the credentials of their staff in a transparent and efficient manner. This can simplify the hiring process, allowing for quicker onboarding and reducing administrative burdens. For example, *ArchiveCore* is a healthcare credentialing startup that utilizes Blockchain technology to speed up background checks for new hires. By validating primary source documents using Blockchain, they offer a more efficient and reliable method for verifying credentials. Another Entity, *Carilion Clinic*, demonstrates the potential cost savings and benefits that can be achieved through the use of Blockchain in credential verification. Overall, Blockchain technology has the potential to revolutionize the credential verification process in healthcare, offering improved efficiency, cost savings, and enhanced trust. As more organizations adopt Blockchain-based solutions for credential verification, the industry can benefit from streamlined processes and increased transparency.

Patient-Centered Electronic Health Records: Blockchain technology can enhance the management and interoperability of EHRs by providing a secure and decentralized system for storing, accessing, and sharing patient data. Patients can have greater control over their health information, granting access to healthcare providers as needed, while maintaining privacy and security. Blockchain-based EHRs can improve data accuracy, reduce duplication of tests and procedures, and enable seamless sharing of medical information across different healthcare organizations. Patient-centered EHRs are a crucial aspect of healthcare, and Blockchain technology can provide significant improvements in managing and sharing medical information. Key benefits and features of using Blockchain for patient-centered EHRs include patient control over medical records, as Blockchain technology enables patients to have control over their own medical records. They can grant access to specific healthcare providers, track who has accessed their records, and maintain the privacy and security of their sensitive information.

As mentioned before, Blockchain provides a decentralized and tamper-proof ledger, ensuring the security and integrity of medical records. The immutability of Blockchain makes it difficult for unauthorized parties to manipulate or alter patient data, thus guaranteeing the accuracy and reliability of the records. As such, patients can manage their data-sharing consent more effectively. They can grant permission for specific healthcare providers or researchers to access their medical information, ensuring that their data is shared only with authorized parties. Blockchain technology can also enable real-time updates to medical records. Patients are notified whenever their information is updated, ensuring that they are aware of any changes made to their records and promoting transparency in healthcare. Time-limited visibility is another important aspect, as Blockchain allows for time-limited visibility of medical records to different parties. Patients can set time restrictions on how long their records are visible to specific healthcare providers, ensuring that access is granted only for the necessary duration. Finally, Companies like *MedicalChain* offer solutions to integrate existing healthcare systems with Blockchain technology. This integration enables medical facilities such as hospitals and clinics to leverage the benefits of Blockchain, including virtual consultations and improved data management. By utilizing Blockchain technology for patient-centered EHRs, healthcare providers can enhance data security, improve patient privacy and control, and facilitate efficient and secure data sharing. The use of Blockchain in healthcare, as demonstrated by companies like MedicalChain, has the potential to revolutionize the way medical records are managed and shared, ultimately improving patient care and outcomes.

Enhanced Security and Commitments Enforced via Smart Contracts: Blockchain technology offers enhanced security for healthcare data by using cryptographic techniques and decentralized consensus mechanisms. Data stored on a blockchain is tamper-resistant and immutable, ensuring the integrity and authenticity of healthcare information. Smart contracts, which are self-executing contracts with predefined rules, can be utilized to automate and enforce commitments, such as consent management, data-sharing agreements, and privacy policies, providing transparency and accountability in healthcare transactions. Indeed, Blockchain technology offers enhanced security for health data storage and sharing, and there are several examples of its implementation in the healthcare industry. For example, *Akiri* operates as a Network as a Service (NAAS) and utilizes Blockchain technology to ensure secure health data sharing. It provides a framework for configuring data layers, verifying data sources and destinations, and enforcing policies in real time. By leveraging Blockchain, Akiri enhances the security and integrity of health data by encrypting it and allowing access only through the use of public keys. *Testd* is another platform that uses Blockchain technology to encrypt patient medical information, particularly for test and vaccine verification. By encrypting the data on the Blockchain, Testd ensures that patient information remains secure and accessible only to authorized parties. Blockchain-based smart contracts can also address issues related to contracts and claims in the healthcare industry. These smart contracts enable parties such as patients, doctors, insurers, and medical companies to authenticate themselves on the network and establish clauses and conditions that are automatically executed and visible to all involved parties. This helps in reducing costs by eliminating the need for intermediaries and streamlining the verification and validation process. Every industry struggles with contract disputes and breach of contract clauses. Seventeen percent of insurance claims are denied for various reasons, for example, duplicate claims, improper registration, and 10% of claims in the medical sector are disputed. Smart contracts can be used to solve these issues. Smart contracts make it possible for healthcare personnel to authenticate themselves on the network and document contract clauses, which is carried out automatically and stays visible to all parties. For example, a patient's visit to a physician will be recorded in the blockchain ledger, and the insurer will be notified of the incident. Smart contracts are a tool that stakeholders can use to validate information and expedite dispute settlements. In short, by utilizing Blockchain technology, these solutions provide increased security, transparency, and efficiency in health data storage, sharing, and contractual processes. The use of encryption, smart contracts, and distributed ledger technology strengthens data protection, reduces fraud, and improves trust among stakeholders in the healthcare ecosystem [13].

Internet of Medical Things (IoMT): The IoMT refers to the network of connected medical devices and sensors that collect and transmit patient data. The IoMT plays a crucial role in the advancement of health and medical information systems. IoMT technology empowers medical devices such as body scanners, wearables, and heart monitors to gather, analyze, and transmit data in real time through the internet. As artificial intelligence (AI) continues to progress, healthcare professionals leveraging the IoMT approach may have the ability to capture an image, detect malignant regions, and even identify concerning cells. This information can then be shared with relevant professionals who have authorized access to the data [14]. Blockchain technology can enhance the security, privacy, and interoperability of IoMT devices by providing a decentralized and trusted infrastructure for data exchange and device management. It enables secure data sharing, real-time monitoring, and automated device authentication contributing to improved patient care, remote monitoring, and precision medicine.

Data Exchange: With the current systems, patients often have limited control over their medical information and face challenges in sharing it with healthcare providers or other relevant parties. Blockchain technology provides a secure and decentralized mechanism for patients to share their medical information, granting them more control and enabling efficient and accurate data exchange. Patients can choose who to share their records with, such as doctors, specialists, or insurance agents, improving collaboration and communication in healthcare.

Data Storage: Blockchain-based healthcare systems utilize decentralized storage on the cloud where transactions are securely saved on blocks. Patient medical information is organized within EHRs and forms part of this decentralized storage system. Cloud storage offers increased capacity, cost-effectiveness, dynamic connectivity, easy accessibility, and rapid distribution of healthcare data. By deploying a healthcare data management system with Blockchain technology on the cloud, privacy measures can be implemented to ensure the accuracy, protection, and transparency of healthcare data.

Electronic Health Records: Traditional paper-based health records pose challenges in tracking and managing patient health information leading to potential errors and delays in treatment. Adopting EHRs overcomes these challenges by providing a digital platform for storing and accessing patient information. EHRs enhance treatment quality, enable quicker provision of services, improve illness management, facilitate preventative care, support better decision-making, and foster collaboration among healthcare providers. The use of Blockchain technology in EHRs further enhances data security, integrity, and patient control over their health information.

Blockchain technology can also address challenges in other areas of healthcare, such as clinical trials, claims and billing, and pharmaceutical companies. In clinical trials, issues arise with personal information privacy, data sharing, and patient enrollment. Blockchain offers secure data exchange, ensuring repeatability and transparency. Smart contracts on a private Ethereum platform can eliminate trust issues and enhance data openness. Additional components built on the blockchain framework, including distributed big data analytics, data management integration, identity management for device privacy protection, and data sharing for collaborative research, further enhance precision medicine and clinical trial capacity.

Claims and billing management involves the critical task of medical billing, which requires effective communication and understanding between healthcare providers, patients, and insurance companies. Blockchain technology can improve this process by implementing a transparent system for all parties involved, addressing issues of excessive billing and lack of trust.

Pharmaceutical companies face challenges related to maintaining medication quality, researching new treatments and ensuring compliance with safety regulations. The lengthy process, including clinical trials, often lacks privacy and security, leading to counterfeit drugs and recalls. Blockchain technology can address these concerns by preserving anonymity, maintaining security through tamper-proof nodes, and facilitating patent protection using private blockchains. Smart contracts ensure transparency, traceability, and integrity, supporting patent protection efforts [14].

As such, the application of blockchain technology in SCM, clinical trials, claims and billing, and pharmaceutical companies provides solutions to challenges in healthcare. By enhancing transparency, security, and authentication, blockchain technology contributes to more efficient and trustworthy healthcare systems [14]. By leveraging Blockchain technology, telemedicine can provide patients with greater autonomy over their information, streamline patient identification processes, and foster improved collaboration among healthcare providers. These benefits contribute to a more efficient and patient-centric telemedicine experience [16].

Artificial intelligence integration

In addition to the benefits of Blockchain technology discussed above, there are several other considerations related to the integration of artificial intelligence (AI) in telemedicine:

Real-time Feedback and Decision Support: AI enables the provision of real-time feedback to medical personnel, allowing them to make more data-driven decisions and improve patient care. Machine learning, natural language processing, and robotics are used to analyze patient data and provide insights that can enhance the quality of care and patient experiences.

Automated Record Analysis: AI, particularly through machine learning algorithms, is employed to analyze vast amounts of medical data, such as health records. This analysis helps identify patterns and trends that can be utilized to customize treatments and enhance patient care. Organizations like *Google*, *IBM*, and *Mayo Clinic* are using AI-based analysis to improve healthcare outcomes.

Remote Patient Monitoring and Engagement: AI facilitates remote patient monitoring (RPM), which involves gathering and transmitting patient information to healthcare providers outside of traditional clinical settings. Connected technologies enable proactive management of chronic diseases and other illnesses, leading to better health outcomes and reduced healthcare costs.

Accurate Patient Diagnosis: AI plays a crucial role in accurate patient diagnosis, utilizing machine learning algorithms to analyze medical data and assist in making precise diagnoses. This has the potential to improve healthcare outcomes and provide timely treatment.

Accessibility and Affordability of Telemedicine: AI advancements contribute to the accessibility and affordability of telemedicine services. Patients can receive necessary healthcare remotely reducing the need for long-distance travel. This is especially beneficial for individuals in rural areas or those who face challenges in accessing traditional healthcare facilities.

Real-time Physician Feedback: AI, through machine learning algorithms, analyzes patient data and offers real-time insights to doctors during consultations. Platforms like Teladoc utilize machine learning to provide physicians with valuable feedback and support during in-person consultations.

Automatic Reminders and Notifications: AI-powered telemedicine services can provide automatic reminders to patients, ensuring they stay on track with their healthcare plans and appointments.

To recap, the combination of telemedicine, AI, and Blockchain offers numerous benefits in healthcare, including improved diagnosis accuracy, enhanced patient care, remote monitoring capabilities, and increased accessibility to healthcare services. These advancements, coupled with the use of Blockchain technology, address critical challenges in healthcare such as data security, interoperability, SCM, and patient-centric care. By leveraging Blockchain, healthcare systems can achieve more efficient, transparent, and patient-centered care. The application of Blockchain technology in data exchange, storage, and EHRs further contributes to the efficiency, security, and patient-centeredness of healthcare systems overall [14, 17].

Alleviating worldwide poverty

As discussed, Blockchain technology has the potential to alleviate poverty by addressing crucial challenges related to identification, healthcare access, and remittances. The lack of identification for parents and children, particularly in underdeveloped countries, perpetuates the cycle of poverty and hinders access to education, healthcare, employment, and assistance programs. Initiatives such as *ID2020*, *BitNation*, and *OneName* leverage blockchain to provide identities for the unidentified, enabling them to access essential services and opportunities.

In developing countries, paper-based medical records pose challenges, especially when individuals are forced to relocate due to economic or political instability. This frequently results in difficulties in tracking vaccination histories, particularly for pediatric vaccines that require multiple doses over specific periods. Blockchain technology can maintain precise records of administered and pending vaccinations, ensuring accurate and timely healthcare interventions.

Improving healthcare for the poor can be facilitated through systems like *CareAi*, which leverages blockchain technology to provide access to primary care services in underdeveloped regions. CareAi aids in diagnosing serious diseases such as malaria, tuberculosis, and typhoid fever in Africa and other countries. The system ensures anonymity of patient information, benefiting undocumented newcomers and individuals who fear deportation while still receiving necessary healthcare services [18].

Additionally, the flow of money through remittances plays a significant role in developing countries' economies. However, substantial amounts are lost annually to fees affecting the funds available for poverty alleviation. By utilizing blockchain technology, intermediaries can be eliminated, leading to reduced remittance costs. This encourages senders to contribute more funds, as the money reaches the intended recipients more efficiently. Moreover, blockchain technology can enhance transparency and accountability in foreign aid, ensuring that funds are allocated appropriately [19].

Through these various applications, blockchain technology has the potential to empower individuals living in poverty by improving access to identification, healthcare, and financial resources, ultimately contributing to poverty reduction efforts worldwide.

According to *Connolly et al.*, several steps can be taken to enhance the accessibility and affordability of telehealth services using Blockchain technology. The first step involves prioritizing primary care and ensuring the best possible virtual patient experience. By establishing a robust primary care system, telehealth services can focus on preventive and chronic condition management, delivering appropriate and cost-effective care. This coordinated approach improves patient health outcomes. The second step is to improve the affordability of telehealth services. By leveraging Blockchain technology, access to care can be provided at more affordable prices without the burden of traditional overhead costs associated with physical clinics and hospitals. This approach aims to prevent the development of costly health conditions through exceptional primary and preventive care. The third step is to ensure widespread access to telemedicine/telehealth services. This can be achieved by offering health plans that include innovative products and provide virtual access to high-quality coverage and care. By integrating telehealth services into health insurance plans, individuals can easily access and utilize these services for their healthcare needs [20].

By following these steps, healthcare providers and policymakers can leverage Blockchain technology to enhance the accessibility and affordability of telehealth services, ultimately improving patient care and health outcomes.

Telemedicine blockchain architectural considerations

When designing the architecture of telemedicine services using Blockchain technology, there are several important considerations to take into account. First, securing databases and protected access is crucial. The Blockchain architecture ensures the security of data by storing new blocks in chronological order, making it practically impossible to alter their contents. Each block contains hash values and timestamps of previous blocks, creating a transparent sequence of events. Any changes made to the data are cross-referenced by all nodes in the system, preventing unauthorized modifications. Multiple approvals are required to update the data and previous versions are stored alongside newer versions allowing for access control and version management.

Another consideration is increasing the range of service providers. Telemedicine services benefit from the availability of experienced personnel and Blockchain-based architectures aim to provide reachability and availability in rural areas. By connecting local healthcare facilities with city hospitals and specialty centers, telemedicine services can expand their service provider network and offer healthcare services to a wider range of patients. Blockchain technology helps ensure secure human-to-machine interactions and maintains data privacy and encryption, which are crucial for the telemedicine system [21].

Telemedicine systems typically follow a hierarchical, tiered structure consisting of three levels. At Level 1, there are local remote telemedicine centers located in primary healthcare facilities in rural and remote areas. Level 2 includes city or district hospitals connected to neighborhood or rural health centers. Level 3 comprises specialty centers attached to city hospitals for additional support in specific diseases. The patient initiates the process by visiting a nearby community health center where a local healthcare provider conducts an initial examination and gathers vital and diagnostic information. This data is then transmitted to the city hospital for review by remote doctors before the live patient interaction begins [22].

The proposed architecture for healthcare using Blockchain technology suggests the use of *Hyperledger Fabric* as a suitable framework. Hyperledger Fabric, an open-source platform hosted by *the Linux Foundation*, offers simplified data sharing, upgraded smart contract technology, and faster transaction speeds. It provides access control and extensive control over smart contracts supporting various programming languages. With transaction speeds of up to 3,000 transactions per second, Hyperledger Fabric enables low transaction costs and high throughput. It offers a private, scalable, and highly configurable infrastructure for managing health records.

Several healthcare projects have leveraged Hyperledger technology to enhance different aspects of the healthcare system, as follows.

KitChain: A blockchain-based pharmaceutical clinical supply application that focuses on providing an immutable record for shipment and event tracking. It eliminates the dependency on paper trails and facilitates collaboration among top pharmaceutical firms within a single supply chain model.

MyClinic.com: A web application that allows patients to view and share their medical data. It enables scheduling appointments, video conferencing with doctors and specialists, and accessing medical results. The application aims to improve patient-physician relationships and expand access to healthcare professionals globally.

Axuall: Utilizes the *Sorvin* network and *Hyperledger Indy* architecture to enable real-time identity and credential verification. It reduces paperwork for doctors and patients and the cryptographic architecture ensures the authenticity of doctors' credentials.

Melloddy: A machine ledger for drug discovery that utilizes machine learning techniques. It leverages Hyperledger Fabric's component *Substra* to ensure traceability, safe distribution of machine learning jobs, and protection of sensitive data. Melloddy aims to improve predicted performance by utilizing all available data while maintaining data privacy.

Verified.me: A digital identification network based on Hyperledger Fabric. It enables users to control access to their data and prevent oversharing by choosing which parties can access their information. Reputed health and wellness providers, such as Dynacare, have joined the Verified.me network.

These projects showcase the diverse applications of Hyperledger technology in healthcare ranging from pharmaceutical supply chains to patient data management and identity verification. They demonstrate how Blockchain technology can enhance security, efficiency, and collaboration in the healthcare industry.

By considering these architectural considerations, telemedicine services can leverage Blockchain technology to enhance data security, increase service provider availability, and improve healthcare delivery to underserved areas [23].

CONCLUSIONS AND RECOMMENDATIONS

To reiterate, Blockchain technology has the potential to revolutionize healthcare by improving data security, efficiency, and patient outcomes. However, several challenges need to be addressed for its widespread adoption. To leverage the benefits of Blockchain in healthcare, the following recommendations should be considered.

Regulatory Compliance: Healthcare organizations must ensure compliance with relevant regulations, such as *HIPAA*, *GDPR*, and *FDA* guidelines. Collaboration with regulatory bodies is crucial to establish clear guidelines for Blockchain implementation in healthcare.

Interoperability: To overcome the challenge of interoperability, industry-wide standards and protocols should be developed. Collaboration among healthcare providers, technology vendors, and standardization organizations can help establish a common framework for data sharing and communication.

Scalability: As the adoption of Blockchain technology grows, scalability becomes critical. Healthcare organizations should explore blockchain scaling solutions, such as shading, sidechain, and layer-two protocols to increase the network's capacity and handle the growing demand.

Collaboration and Piloting: Healthcare organizations should collaborate with technology providers, regulators, and other stakeholders to develop a comprehensive framework for Blockchain implementation in healthcare. Conducting pilot projects in controlled environments can help evaluate the technology's feasibility, effectiveness, and return on investment.

Education and Training: Promoting education and training programs on Blockchain technology for healthcare professionals is essential. Healthcare providers should invest in training their staff to understand the benefits, risks, and best practices associated with Blockchain implementation.

Data Privacy and Consent: Blockchain implementations should prioritize data privacy and consent management. Ensuring that patients have control over their health data and that sensitive information is appropriately protected is crucial.

Continuous Monitoring and Evaluation: Continuous monitoring and evaluation of Blockchain implementations in healthcare are necessary to identify areas of improvement, address challenges, and refine the technology's use cases. Feedback from patients, healthcare providers, and other stakeholders should be actively sought and incorporated into the development process.

Recommended best practices

The provided recommendations for telemedicine best practices in relation to Blockchain technology are comprehensive and cover key aspects of implementation. Here is a summary of the suggested best practices.

Ensure Data Privacy and Security: Implement encryption and access controls to protect patient data stored on the Blockchain.

Verify Data Accuracy and Integrity: Validate and verify data before adding it to the Blockchain to maintain the accuracy and integrity of patient information.

Collaboration with Stakeholders: Involve all relevant stakeholders, including healthcare providers, patients, and regulators in the development and implementation of Blockchain solutions for telemedicine.

Develop Clear Governance and Standards: Establish protocols and standards for data sharing, access control, and data management on the Blockchain.

Address Legal and Regulatory Challenges: Ensure compliance with data protection laws, patient consent requirements, and other relevant regulations when using Blockchain in healthcare.

Start with Small-Scale Pilots: Test and refine the technology through small-scale pilot projects before scaling up to larger implementations.

Limited English Proficiency (LEP): Provide translation services and support for patients with limited English proficiency to ensure effective communication and understanding.

Providing Access to Technology: Ensure that patients, including those with disabilities, have access to the necessary digital tools and resources for telemedicine services.

Overall, these best practices emphasize the importance of data security, collaboration, compliance, and patient accessibility in the implementation of Blockchain technology for telemedicine. By following these guidelines, healthcare organizations can maximize the potential benefits of Blockchain while prioritizing patient privacy and care.

REFERENCES

[1] Telemedicine. (2021, June 11). *Doctors Without Borders / Médecins Sans Frontières (MSF) Canada*. Retrieved October 26, 2022, from https://www.doctorswithoutborders.ca/telemedicine

[2] Team, E. H. (2021, December 17). The evolution and future of telemedicine. *Eden Health*. Retrieved February 13, 2023, from https://www.edenhealth.com/blog/evolution-of-telemedicine/

[3] VSee. (2023, January 18). *What is telemedicine?* Retrieved February 7, 2023, from https://vsee.com/what-is-telemedicine/

[4] Lounds, M. (n.d.). *Blockchain and its Implications for the Insurance Industry*. Munich Re Life US. Retrieved February 7, 2023, from https://www.munichre.com/us-life/en/perspectives/underwriting/blockchain-implications-insurance-industry.html

[5] Takyar, A. (n.d.). *The role of Blockchain in telemedicine technology*. LeewayHertz. Retrieved October 13, 2022, from https://www.leewayhertz.com/blockchain-in-telemedicine/#:~:text=Traceability%20of%20remote%20treatment&text=Moreover%2C%20Blockchain%20enables%20telemedicine%20applications,patients%20to%20propose%20suitable%20treatment

[6] Ahmad, R. W., Salah, K., Jayaraman, R., Yaqoob, I., Ellahham, S., & Omar, M. (2021, April). The role of Blockchain technology in telehealth and telemedicine. *International Journal of Medical Informatics*. Retrieved October 13, 2022, from https://www.ncbi.nlm.nih.gov/pmc/articles/PMC7842132/

[7] Doctors Without Borders / Médecins Sans Frontières (MSF) Canada. (2021, June 11). *Telemedicine*. Retrieved December 8, 2022, from https://www.doctorswithoutborders.ca/telemedicine#:~:text=How%20does%20it%20work%3F,and%20help%20make%20a%20diagnosis.

[8] Doctors Without Borders / Médecins Sans Frontières (MSF) Canada (a). (n.d.). *Innovation: New programs and tools for saving lives*. Innovation: MSF Solutions for Saving Lives. Retrieved October 26, 2022, from https://www.doctorswithoutborders.ca/issues/innovation-new-programs-and-tools-saving-lives

[9] Ivanovska, I. (2022, July 25). *Family doctors advance their knowledge on telemedicine to help build a more resilient primary health care system.* UNICEF global. Retrieved February 13, 2023, from https://www.unicef.org/northmacedonia/press-releases/family-doctors-advance-their-knowledge-telemedicine-help-build-more-resilient

[10] Kurtsikidze, M. (2021, September 22). *EU, UN and Government of Georgia Launch New Digital Health Project.* UNICEF. Retrieved February 13, 2023, from https://www.unicef.org/georgia/press-releases/eu-un-and-government-georgia-launch-new-digital-health-project

[11] de Castro, C. (2022, November 9). *Portal telemedicina: A Population Health Management platform using AI to detect anomalies and prioritize emergencies.* Portal Telemedicina: A population health management platform using AI to detect anomalies and prioritize emergencies | UNICEF Innovation Fund. Retrieved February 13, 2023, from https://www.unicefinnovationfund.org/broadcast/updates/portal-telemedicina-population-health-management-platform-using-ai-detect

[12] Wicklund, E. (2016, October 27). Red Cross turns to telehealth for emergency care. *mHealth-Intelligence.* Retrieved February 13, 2023, from https://mhealthintelligence.com/news/red-cross-turns-to-telehealth-for-emergency-care

[13] Alkhaldi, N. (2022, March 23). *Top blockchain use cases in healthcare, advantages, and challenges.* ITRex. Retrieved October 26, 2022, from https://itrexgroup.com/blog/blockchain-use-cases-in-healthcare-advantages-challenges/

[14] Bruce, J., Zhang, P., & Katyayani, S. K. (2022). The Various Current Usages of Blockchain Technology For Healthcare Applications. In *Top Blockchain Use Cases for Enterprises* (pp. 12–25). Rejolut: Essay.

[15] STL Partners. (2022, November 30). *5 blockchain healthcare use cases.* Retrieved February 7, 2023, from https://stlpartners.com/articles/digital-health/5-blockchain-healthcare-use-cases/

[16] SSIVIX Lab. (2020, August 31). *3 ways how Blockchain Technology is applied in telehealth.* SSIVIX Lab. Retrieved October 13, 2022, from https://ssivixlab.com/3-ways-how-blockchain-technology-is-applied-in-telehealth/

[17] Sun, P. (2022, June 23). *How AI Helps Physicians Improve Telehealth Patient Care in Real-Time.* Arizona Telemedicine Program. Retrieved December 8, 2022, from https://telemedicine.arizona.edu/blog/how-ai-helps-physicians-improve-telehealth-patient-care-real-time

[18] Klein, G. (2019, March 30). *Blockchain Technology and Alleviating Poverty.* The Borgen Project. Retrieved October 25, 2022, from https://borgenproject.org/tag/blockchain-technology-and-poverty/

[19] Yu, R. (2017, March 17). *Three simple ways blockchain can help the poor.* The Borgen Project. Retrieved December 8, 2022, from https://borgenproject.org/ways-blockchain-can-help-the-poor/

[20] Connolly, C., & Hwang, C. (2021, April 15). *Nonprofit Health Plans Launch Telehealth-first options to increase access and affordability.* NEJM Catalyst - Innovations in Care Delivery. Retrieved October 26, 2022, from https://catalyst.nejm.org/doi/full/10.1056/CAT.21.0031

[21] Hossain, C.A., Mohamed, M.A., Zishan, M.S.R. et al. Enhancing the security of E-Health services in Bangladesh using blockchain technology. *International Journal of Information Technology*, 14, 1179–1185 (2022). Retrieved October 26, 2022, from https://doi.org/10.1007/s41870-021-00821-9

[22] Pramanik, P. K., Pareek, G., & Nayyar, A. (2019). Security and privacy in Remote Healthcare. *Telemedicine Technologies*, 201–225. https://doi.org/10.1016/b978-0-12-816948-3.00014-3

[23] 101 Blockchains. (2022, July 15). *How hyperledger is changing the healthcare system.* 101 Blockchains. Retrieved March 5, 2023, from https://101blockchains.com/hyperledger-for-healthcare/

Chapter 11

Non-traditional banking services for the underserved

Ankita Vashisth, Kolawole Salako and Pramitha Pinto

BACKGROUND

According to *CryptoDefinitions*, a significant portion of the global population, approximately 20%, lacks access to financial institutions, leaving around 1.7 to 2.0 billion people without access to a bank [1]. The World Bank's Global Findex report further reveals that about 1.7 billion individuals lack essential banking services. Among this group, half are women who are either unemployed or come from low-income households in rural areas. Additionally, between 2011 and 2017, there was a consistent 9% gender disparity in account ownership in developing countries, impeding women's ability to control their financial affairs [2].

In countries like Mexico, the lack of access to financial services affects 73% of the population, leading to financial and economic marginalization [3]. The majority of those without bank access reside in developing nations, which hampers their ability to contribute to the country's GDP due to their limited financial capacity. One of the primary reasons for this limited access is the remote locations of bank branches. Moreover, unreliable and inadequate customer identification procedures prevent people, particularly those from lower economic strata, from accessing even basic banking services predominantly in developed nations [1].

To address these challenges, blockchain technology emerges as a potential solution [3]. Its inherent characteristics make it suitable for peer-to-peer transactions without involving a third party, enabling faster and more cost-effective transactions that are permanently recorded [1]. The transformative potential of blockchain technology holds promise for billions of people who lack access to banking services [3].

Barriers to accessing non-traditional banking services

Barriers to accessing non-traditional banking services can impede individuals' ability to benefit from these alternative financial options. The following obstacles are commonly encountered:

Non-Standardized Regulations: A major challenge in establishing a competitive market for non-traditional banking services is the lack of clear regulatory requirements. Without defined regulations, it becomes difficult to develop a robust and reliable framework [4].

Lack of Trust in Customers: Many individuals still harbor mistrust toward non-traditional financial service providers, leading them to prefer established large financial institutions. Trust is built over time, and the longstanding reputation of traditional institutions makes customers hesitant to switch to new alternatives [4].

Cyberattacks on Non-Banking Infrastructure Services: Non-traditional financial services heavily rely on information technology systems and electronic data. However, the increasing number of cyber threats, both from external sources and insider threats, pose

DOI: 10.1201/9781003462033-13

Table 11.1 Comparison of transfer charges between western union and cryptocurrency

Countries	Principal amount	Western union international transfer charges		Bitcoin (Binance.com)	Remarks
		Debit/Credit	Cash in store		
India	CAD 990	$10	$7	Trading fees	Trading charges on the
France	CAD 990	$24.80	$7	Maker: 0.02 – 0.10%	Binance exchange are
Kenya	CAD 990	$24.80	$7	Taker: 0.04 – 0.10%	among the lowest in the
Philippines	CAD 990	$4.99	$7		market with the highest
China	CAD 990	$24.80	$7		charge for both maker
					and taker trades being
					0.1%. When a limit order
					to sell an asset is not
					filled quickly a marker
					fee is charged and when
					a market purchase order
					is placed a taker fee is
					charged.

Source: https://www.finder.com/ca/western-union-fees

a significant risk to these systems. Safeguarding against cyberattacks becomes crucial for ensuring the security and integrity of non-traditional banking services [5].

Knowledgeable Employees: Developing countries often face a shortage of skilled employees in the field of financial service technology. The lack of knowledgeable individuals poses a challenge for non-traditional enterprises seeking to provide these services effectively [4].

Corruption: Although blockchain technology is a powerful tool, it alone cannot solve global economic and financial problems. Building wealth requires the efforts of people and corruption within leadership becomes a hindrance to financial inclusion. Addressing corruption is crucial to promoting effective financial services for all [3].

Literacy: A significant challenge in adopting traditional or non-traditional financial services in developing countries is the low literacy rates. For example, in various parts of Africa, only 50% of the population is literate. Literacy plays a vital role in understanding financial concepts, using smartphones, and engaging in online communication, making it an essential aspect of financial inclusion [3].

Cross-Border Payments: Efficient and secure transfer of funds across international borders is a complex process. The involvement of multiple intermediaries and the intricacies of the system make cross-border payments expensive. Streamlining this process is necessary to reduce costs and enhance accessibility [6].

As of August 19, 2021, *Western Union*'s fees for sending an estimated CAD$990 from Canada to various countries for fund transfers were as follows (Table 11.1) [7].

DISCUSSION

Financial inclusion, as highlighted by *Tapscott and Tapscott*, plays a pivotal role in economic development [3]. It refers to the ability of individuals and companies to access affordable and valuable financial products and services [2]. Financial inclusion is crucial for fostering economic growth and alleviating poverty [8].

In the digital era, there are immense opportunities for innovation and economic advancement. Blockchain technology has the potential to unlock the untapped human capital pool, adding billions of dedicated and successful entrepreneurs to the global economy [3]. It enables households to receive assistance in making investments, managing personal finances, and maintaining a balance in consumption [8].

Moreover, blockchain technology facilitates the efforts of both for-profit and nonprofit organizations in addressing social and environmental challenges. *Accenture*, for instance, collaborates with various stakeholders such as businesses, governments, and organizations to ensure the success of social initiatives through the use of blockchain technology. Similarly, *Consensys* blockchain leverages the support of the Ethereum community to assist investors, non-governmental organizations (NGOs), and social entrepreneurs in achieving the sustainable development goals of the world [9].

By harnessing the potential of blockchain technology, financial inclusion can be advanced, promoting economic development and addressing societal issues effectively.

Non-traditional banking activities

Micro-loans, an important form of financial service in some developing nations, provide small loan amounts to small businesses and disadvantaged individuals, who lack access to traditional capital sources. The microfinance industry recognizes the transformative potential of blockchain technology and explores ways to incorporate it [10]. The following are the benefits of utilizing blockchain technology in non-conventional banking activities:

> *Reduced Social-Economic Impact for the Underserved*: Blockchain technology in microfinance breaks down traditional barriers, enabling significant benefits for underprivileged individuals [10]. The availability of financial services has increased in developing nations due to modern payment methods and digitalized banking services [8].
>
> *Wider Reach of Banking Services*: In regions with weak or nonexistent payment networks, blockchain technology can have a profound impact. It empowers the expansion of various projects and microcredit businesses globally. By leveraging blockchain, companies in financial services, mobile, and other sectors can tap into the economic potential of the financial industry [3].
>
> *Cost Reduction and Transparency of Transactions*: Blockchain technology can mitigate expensive operational management costs, bringing substantial benefits to the underprivileged and underserved [9]. Its decentralized nature eliminates the need for intermediaries or trust authorities in microfinance institutions. Consequently, the interest paid by individuals in exchange for financial services can be significantly reduced [10].

By harnessing the capabilities of blockchain technology, microfinance institutions can enhance their operations, expand access to financial services, and promote financial inclusion for disadvantaged individuals and small businesses. The cost savings, increased transparency, and wider reach of services contribute to the empowerment and economic upliftment of individuals, families, and small businesses in developing economies.

Non-traditional banking services initiatives in Costa Rica

Latin American countries, such as Argentina, Brazil, Colombia, and Mexico, require improvements in access to financial services. The lack of access to these services can negatively impact people's living standards and hinder the development of new businesses.

According to the Central Bank of Costa Rica, there has been a significant increase in mobile phone usage with internet services, with approximately 90% of the population using mobile phones [8]. Between 2014 and 2017, there was a reported 20% increase in registered mobile phone lines per 100 residents. In Costa Rica, the electronic payment system and digital banking have achieved high standards. Over the past decade, the national payments system (*SINPE*) has facilitated a growing number of mobile and digital banking transactions. SINPE offers seamless currency transfers in *Colón*, Dollar, and Euro, providing convenient and efficient payment solutions. Additionally, contactless bank cards handle 50% of all non-cash transactions, further enhancing convenience and accessibility [8].

An important initiative in Costa Rica is the conversion of over 14,000 prepaid beneficiary accounts to debit card accounts through the *Avancemos* program. Avancemos is the country's largest conditional cash transfer program, and this conversion was made possible by the *Institute for Social Assistance* (IMAS) [8]. Including more indigenous groups, such as *Bribri*, in this program can help reduce the number of communities experiencing high rates of financial exclusion. By expanding the reach of the program, more individuals can benefit from financial services and improve their financial well-being.

Improving access to financial services in Latin American nations requires efforts to enhance digital infrastructure, promote electronic payment systems, and ensure inclusive programs that cater to diverse communities. By addressing these challenges, individuals and businesses can thrive, leading to overall economic growth and development in the region.

Microloan financing programs

Blockchain technology is seen as a potential solution for microfinance institutions to fulfill their promises of financial development. Its ability to eliminate intermediaries enables economic autonomy for individuals who have limited access to traditional financial services. In Brazil, where one-third of the population faces restricted financial access and high interest rates, blockchain-based startup *Moeda* collaborates with local credit cooperatives to extend affordable services to rural areas [11].

One of blockchain's key advantages, as highlighted by Scott Nelson, CEO of *Sweet Bridge*, is its ability to leverage existing affinity networks to establish and verify borrower identities [11]. This is particularly significant as approximately 1.7 billion individuals worldwide lack access to bank accounts due to inadequate identification [12]. Blockchain is considered a preferred technology to address identity concerns and provide individuals with the necessary digital identification, also known as digital ID, to access essential services such as banking, government benefits, and education [13].

Designing effective digital ID programs requires careful consideration of infrastructure, governance, and compliance with regulations. Data privacy policies are essential to ensure individuals' control over their data and the responsible handling of it by institutions. Digital ID can be used to comply with regulations that require identification for financial services, but efforts must be made to avoid excluding individuals without digital IDs or those who choose not to use them [14].

Governments around the world are striving to integrate their populations into the formal economy through various digital initiatives. For example, China relies on domestic technology companies and smartphone applications linked to financial institution accounts for financial transactions. India has successfully implemented the world's largest digital biometric ID program, and Kenya is renowned for its mobile money model, providing customized financial services through mobile accounts [13].

The *Kiva Protocol*, a collaboration between the Sierra Leonean government and microloan startup Kiva, serves as an example of a public–private partnership that addresses barriers to financial access. The protocol establishes a digital infrastructure that allows secure data exchange between financial institutions, clients, and government agencies. It addresses the lack of legal identification and verified credit history, enabling Sierra Leoneans to open bank accounts using their thumbprints and integrating biometric and distributed ledger technologies. The program aims to encourage banks to provide loans to individuals without credit history who were previously overlooked [12, 15].

By leveraging blockchain technology and digital ID systems, microfinance institutions and governments can overcome barriers to financial inclusion, empowering individuals and fostering economic development. These initiatives demonstrate the potential of blockchain to revolutionize access to financial services, particularly for underserved populations.

BanQu is a pioneering solution that utilizes blockchain technology to create an economic identity and promote financial inclusion. It provides a digital identity for individuals without official credentials or credit records. Users can create an account on BanQu using any mobile phone and in any language. Transactions made on the platform are recorded on the blockchain, ensuring security and immutability. Proof of credibility, such as land ownership, vaccine records, remittance income, and microloan history can be integrated to access microloans [16]. The decentralized nature of blockchain ensures that all transaction partners have equal access to and ownership of personal data, contributing to the formation of an individual's economic identity [16].

The government of India introduced *Aadhaar*, a digital ID program that assigns a 12-digit personal identification number to individuals and connects it with their fingerprints and iris scans. Aadhaar is similar to the Kiva Protocol and aims to provide a reliable identification system [12].

Colendi is another blockchain-based microcredit platform that uses various data sources, including smartphone data, social media activity, and transaction histories to generate accurate credit scores and provide inclusive financial services to unbanked individuals and businesses. Colendi ID and Colendi Score are the two main products of the platform, and the decentralized data storage on the Ethereum blockchain ensures transparency and privacy [17].

AID Tech is an organization that leverages blockchain technology and digital ID platforms to facilitate quick and easy payments for low-income minorities in areas such as aid, contributions, e-health, disaster relief, remittances, and social welfare. Their platform has been successfully implemented in countries like Uganda, Egypt, Morocco, Jordan, and India [18].

These examples highlight the transformative potential of blockchain technology and digital ID platforms in promoting financial inclusion and providing alternative financial services to underserved populations. By leveraging decentralized technologies, these initiatives aim to overcome barriers and empower individuals with greater access to financial opportunities.

Non-traditional services inclusiveness using digital platforms

The rise of non-traditional financial service providers, such as FinTech companies, has brought about significant changes in the banking industry and the way banks interact with their customers. These alternatives to traditional banks aim to simplify the process of accessing loan and savings services for clients [4, 19].

One notable example of the impact of digital platforms on financial services is the mobile money revolution in Sub-Saharan Africa, with Kenya's *M-PESA* being a prominent case study. M-PESA, launched by *Safaricom* in 2007, is a mobile wallet and payment platform that connects unbanked customers to formal financial services in collaboration with banks, NGOs, and microcredit organizations. It enables customers to send and receive payments cheaply and securely through text messages and provides them with the ability to save money and withdraw it from various retail locations across the country [20].

The use of digital platforms to reach individuals without access to traditional banking services offers several benefits:

Wider Reach to the Underserved: Contemporary payment methods and digital banking services have increased the availability of financial services in OECD nations. With the integration of blockchain technology, projects, and microcredit businesses will expand further, empowering individuals without standard banking services [3, 21].

Reduction of Transaction Administrative Costs: Unlike traditional banks that often have high administrative costs, blockchain technology allows for direct peer-to-peer transactions without the need for intermediaries. This enables instant movement of money across borders or into areas affected by natural disasters, significantly reducing administrative expenses [9, 22].

Mitigation of Risks and Fraud: Data recorded on a blockchain is digitally stored and accessible to all network participants, reducing the possibility of fraudulent activities or transaction duplication. By incorporating microloans on the blockchain and granting clients access, businesses can be held more accountable for their actions, enhancing transparency and accountability [3].

These advantages highlight how digital platforms and technologies, including blockchain, can revolutionize financial services and enable greater financial inclusion for underserved populations while reducing costs and enhancing security.

Benefits of blockchain in financial services

Blockchain technology has the potential to bring significant benefits to the financial services sector. The following are some key advantages of using blockchain in financial services:

Cost and Time Efficiency: Blockchain enables efficient and cost-effective trading and settlements in the financial industry. With instant transaction finality and the elimination of middlemen, banks can save time and reduce operational expenses. Automated settlements through smart contracts can further enhance efficiency, automating processes based on predefined criteria [20, 23].

Improved Transparency and Accountability: The use of blockchain provides greater visibility among financial firms, regulators, and central banks. Authorities can have access to the blockchain, leading to better regulatory reporting and monitoring. Additionally, blockchain's decentralized nature reduces the risk of compromising credit risk score information as customer data is managed through decentralized programs like smart contracts. This enhances transparency and accountability in credit rating and loan decisions [24].

Enhanced Security and Data Protection: Decentralized storage of customer records in the blockchain reduces the vulnerabilities associated with centralized storage and improves data security. Once recorded on the blockchain, customer information becomes irreversible, ensuring data integrity. This can help mitigate data security and cybercrime concerns [25].

Facilitating Microcredit and Financial Inclusion: Blockchain technology enables improved communication between banks and microcredit organizations. Instead of relying solely on credit scores, lenders and providers can track the use and repayment of microcredits on the blockchain. This decentralized approach reduces credit risk score vulnerabilities and enables more inclusive access to financial services [3].

Transparency in Charitable Initiatives: Blockchain-based financial services can enhance transparency and accountability in charitable initiatives. Donations recorded on the blockchain are immutable, ensuring permanent and transparent records for donors. This helps build trust and confidence in charitable organizations as the use of funds can be tracked, preventing mismanagement or misuse of donations [3, 26].

The following are some examples of charitable initiatives that leverage blockchain-based financial services:

World Food Programme (WFP): WFP, a UN initiative, utilizes blockchain technology to provide cash transfers to refugees in conflict zones. Through their Building Blocks system, funds are distributed directly to program participants via mobile payments, reducing costs and eliminating middlemen.

AIDChain: AIDChain is a charity platform that connects charitable organizations, donors, recipients, and service providers. It allows charities to promote projects and collect cryptocurrency donations. AIDChain supports multiple currencies and facilitates trading of crypto for cash for beneficiaries.

Binance Charity: Binance Charity Foundation, the charity arm of the Binance cryptocurrency exchange, supports education, disaster assistance, and rebuilding efforts. Blockchain technology is used to enhance contribution transparency, allowing donors to choose projects, contribute through the Binance Charity Wallet, and track the impact of their donations.

BitGive Donation Project: BitGive is a blockchain organization that launched *GiveTrack*, a blockchain-based contribution platform. GiveTrack provides transparency by recording data on the blockchain at each stage of the donation process. Donors can view real-time financial statistics and outcomes of the initiatives they have sponsored [27].

These initiatives leverage blockchain to increase transparency, accountability, and efficiency in charitable giving, ensuring that funds are allocated and deployed effectively while providing donors with a clear view of the impact of their contributions.

Blockchain framework for non-traditional banking services

Hyperledger, Ethereum, and *Corda* are prominent blockchain systems, each with its own unique characteristics and areas of application.

Hyperledger is an open-source project hosted by the *Linux Foundation*. It focuses on developing enterprise-grade blockchain solutions for various industries. Hyperledger provides a modular framework that supports interoperability and compatibility across different projects. It aims to facilitate the development of industrial blockchain applications and offers tools, libraries, and frameworks for building distributed ledger systems. *Hyperledger Fabric* and *Hyperledger Sawtooth* are two well-known platforms within the Hyperledger ecosystem.

Ethereum is a decentralized, open blockchain platform that supports the creation of smart contracts and decentralized applications (DApps). It is known for its programmable blockchain, allowing developers to build and deploy their own applications on the Ethereum network. Ethereum uses its native cryptocurrency called *Ether* (ETH) to incentivize participants and execute smart contracts. It has gained popularity for its versatility and wide range of potential use cases beyond just financial applications.

Corda is a blockchain platform developed by *R3*, a distributed database company. Corda is designed specifically for the financial services industry, focusing on managing legal contracts and shared data between trusted parties. It emphasizes privacy and confidentiality, enabling secure transactions and communication among multiple applications on a single network. Corda uses a "notary" mechanism to validate transactions and achieve consensus.

While Ethereum provides a more general-purpose blockchain platform for various decentralized applications, Hyperledger focuses on enterprise use cases, and Corda specifically caters to the needs of the financial services sector. Each platform has its own strengths and features that make it suitable for different scenarios and industries [28].

Corda blockchain architecture

As mentioned, Corda, developed by R3, is an open-source, permissioned blockchain platform designed for commercial use in industries with stringent regulations [29]. Combining characteristics of both blockchain and distributed ledger technology (DLT), Corda aims to reduce the effort required to maintain data integrity, scale efficiently, and handle a high volume of transactions [30]. Unlike public blockchains, which lack privacy and may not meet the needs of businesses and banking services, Corda provides a viable option with its emphasis on transaction privacy and known participants [30]. Permissioned blockchains like Corda offer several advantages over permissionless blockchains, including reduced computational requirements, predetermined validating nodes, quicker rule updates, and cost-effectiveness for businesses [31, 32].

The conceptual architecture of Corda sets it apart from other ledger platforms, addressing issues of scalability and privacy. Corda enables simultaneous processing of multiple transactions while meeting strict privacy requirements [33]. The following are some key framework concepts within Corda.

Privacy is a crucial element in a dependable digital ledger system. Corda allows parties to enter transaction details while ensuring data protection through dispersion among various servers and nodes. Multiple participants can transact securely by exchanging relevant information and Corda's advanced privacy features safeguard the interests of businesses and organizations without compromising security [34].

Establishing a closed network among known participants is essential for a permissioned blockchain system like Corda. Corda ensures that network participants can reliably identify each other, assigning a single user profile to legal entities such as companies or individuals based on their identity structure [35].

Consensus mechanisms play a vital role in enabling organizations on a decentralized network to agree on transactions and maintain the integrity of the blockchain. Corda employs consensus techniques like Byzantine Fault Tolerant algorithms to verify transactions. Its flexible architecture supports multiple consensus pools using various algorithms, providing users with a customizable consensus model based on their requirements [31].

Distributed systems based on blockchain involve business logic, rule validation, smart contracts, and program files. In Corda, proposed alterations to the ledger are verified by having each participant execute the same code, ensuring transparency and consistency in transaction processing [35].

Smart contracts are blockchain-based applications that execute specific events when predefined conditions are met [26]. Corda utilizes smart contracts for its business logic, keeping them simple and reusable. Users can create agreements using basic reusable functions, accept or reject proposed contracts, and the contract code governs the validity of transactions. Corda follows a model similar to Bitcoin, where a transaction is proposed by one party and verified by other network users. Java Virtual Machine (JVM) is chosen for executing and verifying smart contracts, leveraging its extensive libraries and bytecode set. This allows users to use their preferred programming language, simplifying the application of contract codes and promoting positive development [36].

In philanthropy, smart contracts can automate and streamline the contribution process between contributors and charity organizations. These contracts can include terms and conditions specifying when and where funds, supplies, or items should be transferred. They can also regulate the allocation of funds toward supporting charitable projects, simplifying the entire gift distribution process from collection to recipient selection and finalizing transactions with beneficiaries [26].

Corda use cases

Corda's primary objective is to provide users with a platform that offers standard services and ensures compatibility with network members. By enabling organizations to trade via smart contracts, Corda aims to overcome obstacles that hinder commercial transactions. Its use cases adhere to strict privacy and security standards, making it increasingly popular in the financial industry. In terms of programming languages, Corda differentiates itself from Hyperledger. While Hyperledger is built on Google's Golang language, Corda utilizes *Kotlin*, a programming language targeting *JavaScript* and the JVM. By leveraging Kotlin, Corda achieves seamless integration with other programming paradigms, making it superior to Hyperledger for financial industry applications [37].

Corda blockchain technology finds application in various institutions through the following use cases:

Finance: The financial industry has been an early adopter of blockchain technology. Corda enables the creation of a streamlined channel for fast transactions and secure customer data. It enhances the network by improving operational efficiencies, reducing costs, and speeding up processing times. The immutability and transparency of data foster increased institutional trust. In the payment industry, Corda blockchain can facilitate faster transactions, especially in emergency situations [38].

Digital Identity: Managing digital identity is another significant use case for R3 Corda. Businesses need to ensure proper implementation of personal data security. The current strategy often suffers from poor data management and the sharing of personal information. Corda provides a robust solution by adopting DLT/blockchain for digital identity management. It handles tasks such as verification, data collection, and lifecycle management, ensuring smooth operations and error-free processes. Corda keeps sensitive personal data off-chain and offers a secure distributed data network. It is also compatible with self-sovereign identity and corporate Know Your Customer (KYC) practices [38].

One notable example of Corda's adoption is by the *Hongkong and Shanghai Banking Corporation* (HSBC), one of the world's largest banks. HSBC became the first financial institution to migrate the Corda Enterprise blockchain platform, developed by R3, to the Google Cloud. This transition significantly reduces costs and shortens client onboarding times from months to weeks. HSBC employs Corda technology for its Digital Vault, a custody blockchain platform. The bank aims to utilize blockchain to custody digital asset classes in the future. By leveraging Corda technology on the Google Cloud, HSBC gains the ability to move more transaction lifecycle processes onto the ledger, including issuing digital tokens instead of paper certificates. The Digital Vault service, which went live in November 2019, digitizes transaction records of private placement assets like equity, debt, and real estate. This enables global custody clients to access real-time details about their private assets and facilitates audits of transaction records, including tracking coupon receipts for private debt transactions [39].

CONCLUSIONS AND RECOMMENDATIONS

The chapter's exploration of non-traditional banking services for the underserved emphasizes the significance of financial inclusion as a cornerstone of economic development. It highlights that both economic growth and poverty alleviation depend on ensuring access to basic financial services. The lack of such access poses a major challenge, leading to financial and economic marginalization. This issue predominantly affects individuals in developing countries who are unable to contribute to their country's GDP due to limited financial capacity. Barriers to accessing non-traditional banking services include non-standardized regulatory requirements, consumer mistrust of non-traditional financial institutions, prevalent corruption, cross-border payment challenges, and low financial literacy in developing countries. However, the digital era presents vast opportunities for innovation and economic development for the underserved, with blockchain technology holding immense potential. Blockchain can reduce the socioeconomic impact on people without access to banking services in regions with weak or nonexistent payment networks. It can empower the underprivileged by eliminating the need for intermediaries and providing economic autonomy where traditional financial institutions fail to serve them. Blockchain is particularly well-suited to address identity concerns among these individuals. Well-designed digital ID programs, supported by robust infrastructure and effective governance, can overcome obstacles related to lack of legal identification and verified credit history, enabling access to financial services. Utilizing digital platforms to reach those without standard banking services offers a wider reach to the underserved population.

Blockchain-based financial services also have applications in charitable initiatives. Donors are increasingly drawn to charitable organizations that leverage blockchain technology for enhanced accountability and transparency. Prominent blockchain systems like Hyperledger, Ethereum, and Corda have demonstrated the benefits and uses of distributed ledger technology. Among these platforms, Corda, a private and permissioned service, finds significant utility in the financial services sector. Its conceptual design sets it apart from other ledger platforms, addressing risks such as scalability and privacy. Corda's network can handle multiple transactions simultaneously while meeting stringent privacy requirements. By enabling organizations to trade via smart contracts, Corda aims to overcome challenges in commercial transactions. In the realm of philanthropy, smart contracts can automate and streamline the contribution process, offering numerous possibilities for all involved parties.

While there is still a significant number of unbanked individuals worldwide, the trend has been decreasing in recent years. Financial institutions are developing alternative solutions to reach those without access to banking services. It is crucial to focus on the distribution and design of basic financial services to address financial exclusion effectively. Creating and promoting accessible financial services is vital for removing barriers faced by excluded citizens who are most vulnerable to financial exclusion. Resolving these challenges is necessary to achieve financial inclusion and foster economic development.

Suggested good practices

To address the issues of financial exclusion and enhance financial inclusion, several factors and strategies can be considered:

Remove Adoption Barriers: Eliminating barriers to accessing traditional banking services is crucial when targeting the unbanked population. This includes removing monthly account fees and minimum balance requirements that may deter individuals from opening accounts.

Reduce Reliance on Physical Branches: Emphasize mobile banking as a primary channel for accessing financial services. By implementing biometrics or two-factor authentication for identity verification, financial institutions can enable individuals to open accounts and access services online, reducing the need for in-person visits.

Increase Credit and Loan Options: To assess creditworthiness, consider incorporating non-traditional payment data such as timely rent and utility bill payments. Expanding the availability of credit-building loan options, such as secured credit cards and credit-builder personal loans, can also help individuals build credit history. Additionally, explore peer-to-peer lending or crowdfunding platforms to provide financial services to traditionally underserved customers.

Adopt Blockchain for Payments: Blockchain technology can significantly improve cross-border payments by enabling faster, more secure, and cost-effective transactions. Traditional banking channels often involve lengthy processing times and complex fee structures. Implementing blockchain solutions can simplify and expedite the payment process, reducing costs, and enhancing security.

Develop Effective and Secure Mobile Applications: Focus on creating mobile banking applications with extensive functionality to provide users with access to a range of financial services. Such applications should enable users to transfer funds between accounts, pay bills, and even deposit paper checks, maximizing convenience and accessibility.

Expand Digital ID Technology: Implement digital ID initiatives with robust governance measures to ensure the security and transparency of user data. Decision rights, access rights, enforcement methods, and contingency planning are essential aspects of effective governance. Blockchain technology can decentralize information storage, enhancing security, and resilience against cyberattacks or internal fraud.

Advocate for Favorable Regulations: Regulatory frameworks play a crucial role in enabling innovations and addressing financial exclusion. Simplifying regulations and creating an environment that encourages private businesses, especially non-traditional financial service providers, to develop innovative products can help reach the unbanked population. Adapting or streamlining KYC and Anti-Money Laundering (AML) laws can also facilitate access to financial services for underserved individuals.

Incorporate Transparency and Accountability in Charitable Initiatives: Blockchain technology can enhance transparency, accountability, and efficiency in charitable initiatives. By implementing blockchain-based charity administration platforms, organizations can create auditable and transparent systems for collecting and distributing funds. This can help build donors' trust and confidence by ensuring clear evidence of achievement and reducing administrative costs through automation and crowdfunding concepts [40, 41].

By considering these factors and strategies, financial institutions and organizations can work toward achieving financial inclusion and addressing the needs of the underserved population.

REFERENCES

[1] CryptoDefinitions. (2020, August 11). *How is blockchain technology being used to bank the unbanked*. Retrieved October 12, 2022, from https://cryptodefinitions.com/how-is-blockchain-technology-being-used-to-bank-the-unbanked/

[2] The World Bank. (2022, March 29). *Financial Inclusion: Financial inclusion is a key enabler to reducing poverty and boosting prosperity*. World Bank. Retrieved January 22, 2023, from https://www.worldbank.org/en/topic/financialinclusion/overview#1

[3] Tapscott, Don, and Alex Tapscott. *Blockchain Revolution: How the Technology Behind Bitcoin Is Changing Money, Business, and the World, Penguin Canada*, 2016. ProQuest Ebook Central. https://ebookcentral.proquest.com/lib/concordiaab-ebooks/detail.action?docID=6815840.

[4] Kasradze, T. (2021, October). *Emergence of Non-Traditional Financial Service Providers in the Market -A Threat or An Opportunity for the Georgian Banking Industry*. Retrieved October 12, 2022, from https://www.researchgate.net/publication/355926072_Emergence_of_Non-Traditional_Financial_Service_Providers_in_the_Market_-A_Threat_or_An_Opportunity_for_the_Georgian_Banking_Industry

[5] U.S. GAO. (2022, October 13). *Protecting critical infrastructure from cyberattacks*. gao.gov. Retrieved November 14, 2022, from https://www.gao.gov/blog/protecting-critical-infrastructure-cyberattacks

[6] Precisa. (2022, January 24). *Cross-border payments: Challenges & Trends in 2022*. Precisa. in. Retrieved December 4, 2022, from https://www.precisa.in/blog/cross-border-payments-challenges-trends-2022/

[7] Anna Kierstan . (2021, August 19). *How much are Western Union Fees in Canada: Finder Canada*. finder CA. Retrieved March 26, 2023, from https://www.finder.com/ca/western-union-fees

[8] OECD iLibrary. (2020). *Boosting access to credit and ensuring financial inclusion for all | OECD Economic Surveys: Costa Rica 2020*. OECD iLibrary. Retrieved October 11, 2022, from https://www.oecd-ilibrary.org/

[9] Sharma, T. K. (2020, July 20). *Blockchain role in Social Impact Initiatives*. Blockchain Council. Retrieved October 12, 2022, from https://www.blockchain-council.org/blockchain/blockchain-role-in-social-impact-initiatives/

[10] Joshi, N. (2022, March 16). *Empowering microfinance with Blockchain*. bbntimes.com. Retrieved November 15, 2022, from https://www.bbntimes.com/technology/empowering-microfinance-with-blockchain

[11] Stanley, A. (2021, September 13). *Microlending startups look to blockchain for loans*. CoinDesk Latest Headlines RSS. Retrieved December 4, 2022, from https://www.coindesk.com/markets/2017/12/08/microlending-startups-look-to-blockchain-for-loans/

[12] Ayala, T. (2021, September 8). *Banking the unbanked: Africa's biometric blockchain*. SocialFintech. org. Retrieved February 4, 2023, from https://socialfintech.org/banking-the-unbanked-africas-biometric-blockchain/

[13] International Finance. (2021, July 30). *Digital identity acting as a catalyst for financial services*. Internationalfinance.com. Retrieved February 6, 2023, from https://internationalfinance.com/digital-identity-acting-as-a-catalyst-for-financial-services/

[14] White, O., Madgavkar, A., Manyika, J., Mahajan, D., Bughin, J., McCarthy, M., & Sperling, O. (2019, April 17). *Digital identification: A key to inclusive growth*. www.mckinsey.com. Retrieved February 25, 2023, from https://www.mckinsey.com/capabilities/mckinsey-digital/our-insights/digital-identification-a-key-to-inclusive-growth

[15] Cheney, C. (2019, August 21). *In Sierra Leone, New Kiva Protocol uses Blockchain to benefit unbanked*. devex.com. Retrieved February 4, 2023, from https://www.devex.com/news/in-sierra-leone-new-kiva-protocol-uses-blockchain-to-benefit-unbanked-95490

[16] Global Opportunity Explorer. (2019, May 6). *Building an economic identity on Blockchain*. goexplorer.org. Retrieved February 4, 2023, from https://goexplorer.org/building-an-economic-identity-on-blockchain/

[17] Colendi. (2019, May 10). *Financial inclusion without the traditional banking system*. medium. com. Retrieved February 25, 2023, from https://medium.com/colendi/financial-inclusion-without-the-traditional-banking-system-7f067041ef2d

[18] The Million Lives Club. (2022, January 25). *Digital identity platform harnessing blockchain and AI to transform access to social and financial services for the underserved - million lives collective*. The Million Lives Club. Retrieved February 6, 2023, from https://www.millionlives.co/members/aidtech

[19] Raspa, S. (2020, April 22). *6 strategies for banking the unbanked*. Finextra Research. Retrieved February 25, 2023, from https://www.finextra.com/blogposting/18674/6-strategies-for-banking-the-unbanked

[20] Burns, S. (2020, February 29). *Banking the unbanked: Lessons from the developing world*. AIER. Retrieved February 25, 2023, from https://www.aier.org/article/banking-the-unbanked-lessons-from-the-developing-world/

[21] Jérusalmy, O., Fox, P., Hercelin, N., & Mao, L. (2020, July 7). *Basic Financial Services - Finance Watch*. Retrieved February 25, 2023, from https://www.finance-watch.org/wp-content/uploads/2020/07/basic-financial-services-report-fw-july-2020.pdf

[22] Anwar, H. (2022, August 15). *Blockchain in payment: Accelerating payment services*. 101 Blockchains. Retrieved March 1, 2023, from https://101blockchains.com/blockchain-in-payment/

[23] Unbanked. (2022, July 27). *5 common benefits that blockchain has on financial services*. unbanked.com. Retrieved February 11, 2023, from https://unbanked.com/5-common-benefits-that-blockchain-has-on-financial-services/

[24] Rijmenam, M. V. (2020, September 4). *Blockchain - 7 benefits for the financial industry - Mark Van Rijmenam*. londonspeakerbureau.com. Retrieved February 11, 2023, from https://londonspeakerbureau.com/blockchain-7-benefits-financial-industry/

[25] Rejolut. (2023, January 17). *Blockchain in financial services: Usage benefits*. www.Rejolut.com. Retrieved January 28, 2023, from https://rejolut.com/blog/blockchain-in-financial-services

[26] PixelPlex. (2020, November 24). *Blockchain in Charity & Philanthropy: Benefits and Future*. PixelPlex.io. Retrieved November 15, 2022, from https://pixelplex

[27] EconoTimes. (2016, December 9). *BitGive launches blockchain-based real-time donation tracking system 'GiveTrack'*. econotimes.com. Retrieved February 12, 2023, from https://www.econotimes.com/BitGive-launches-blockchain-based-real-time-donation-tracking-system-GiveTrack-442788

[28] Aghili, S., Kaur, M., & Lnu, N. (2022, November 3). *The auditor's guide to blockchain technology: Architecture, use cases*. Taylor & Francis. Retrieved February 4, 2023, from https://doi.org/10.1201/9781003211723

[29] Yafi, R. E. (2022, August 11). *Notaries in Corda 5: An overview: Corda blog*. Corda. Retrieved February 11, 2023, from https://corda.net/blog/notaries-in-corda-5-an-overview/

[30] Rehman, A. U. (2023, January 20). *What is Corda Blockchain? how it helps businesses?* Crypto Economy. Retrieved February 11, 2023, from https://crypto-economy.com/corda-blockchain/

[31] Newton, D. (2018, June 5). *What is Corda?* Corda. Retrieved February 11, 2023, from https://corda.net/blog/what-is-corda/

[32] 101 Blockchains. (2022, August 15). *Permissioned vs permissionless blockchains*. 101blockchains.com. Retrieved February 16, 2023, from https://101blockchains.com/permissioned-vs-permissionless-blockchains/

[33] Matsuzaki, T. (2020, March 24). *Architecture and overview of R3 Corda*. tsmatz. Retrieved February 11, 2023, from https://tsmatz.wordpress.com/2020/03/23/corda-tutorial-on-azure/

[34] 101blockchains. (2021, March 2). *What is Corda Blockchain? simply explained - PUBLISH0X*. Retrieved February 11, 2023, from https://www.publish0x.com/101blockchains/what-is-corda-blockchain-simply-explained-xwqepqo

[35] Xoriant Marketing. (2021, July 22). *Corda – a blockchain framework with no chain of blocks*. Xoriant. Retrieved February 11, 2023, from https://www.xoriant.com/blog/corda-a-blockchain-framework-with-no-chain-of-blocks

[36] Anwar, H. (2022, August 15). *Corda blockchain: Ruler of the financial enterprises*. 101 Blockchains. Retrieved February 11, 2023, from https://101blockchains.com/corda-blockchain/

[37] Geroni, D. (2022, December 6). *Hyperledger vs Corda vs Ethereum: The ultimate comparison*. 101blockchains.com. Retrieved February 4, 2023, from https://101blockchains.com/hyperledger-vs-corda-r3-vs-ethereum/

[38] 101 Blockchains. (2022, August 15). *Top 10 Corda use cases you should know about*. 101blockchains.com. Retrieved March 2, 2023, from https://101blockchains.com/corda-use-cases/

[39] Global Cloud Platforms. (2021, March 7). *HSBC becomes first financial institution to move Corda Enterprise Blockchain technology on to google cloud*. globalcloudplatforms.com. Retrieved March 4, 2023, from https://globalcloudplatforms.com/2021/03/07/hsbc-becomes-first-financial-institution-to-move-corda-enterprise-blockchain-technology-on-to-google-cloud/

[40] Farooq, M. S., Khan, M., & Abid, A. (2020, February 21). A framework to make charity collection transparent and auditable using blockchain technology. *Computers & Electrical Engineering.* Retrieved March 4, 2023, from https://doi.org/10.1016/j.compeleceng.2020.106588

[41] Rangone, A., & Busolli, L. (2021, March 3). *Managing charity 4.0 with Blockchain: A case study at the time of Covid-19 - International Review on public and nonprofit marketing.* SpringerLink. Retrieved March 4, 2023, from https://link.springer.com/article/10.1007/s12208-021-00281-8

Chapter 12

Use of blockchain in Islamic finance

Aditya Jayeshkumar Bhatt and Kesha Sisodia

BACKGROUND

Financial Technology, commonly known as *Fintech*, refers to the use of technology to provide innovative financial solutions. The term originated in the early 1990s with the Financial Services Technology Consortium, a collaborative project by *Citigroup* aimed at advancing technological capabilities [1]. In addition to established financial institutions, Fintech encompasses emerging tech-based startups that adapt to evolving customer needs and preferences by offering creative financial solutions. Collaboration between traditional institutions and startups in the Fintech sector promotes innovation, improves the user experience, and enhances mobility throughout the financial service value chain [2].

The popularity of Fintech has been steadily increasing due to its cost-effectiveness, transparency, ability to provide robust security, speed, and accessibility [3]. A pivotal moment in the development of Fintech was the global financial crisis of 2008. This crisis eroded public trust in the traditional banking system, leading individuals to seek alternative options that offered greater security and confidence in their investments. Moreover, the demand for financial services that are more affordable, flexible, and expedient has been a driving force behind the growth of Fintech [4].

The fintech evolution

The evolution of Fintech can be categorized into three significant eras, as outlined in Table 12.1.

DISCUSSION

Islamic finance basics

Islamic finance refers to a form of financing that adheres to *Sharia* law, which is based on Islamic principles. It encompasses investments and financial activities that comply with Islamic guidelines. Islamic Fintech, specifically, holds great potential in Muslim nations, where the advancement of Fintech has been facilitated by the widespread use of mobile and smartphone technology. However, there are challenges associated with harnessing these opportunities. The primary concerns for Islamic Fintech companies revolve around regulatory frameworks and the availability of high-quality and reliable research to ensure compliance with Sharia principles [4].

The market for Islamic finance is estimated at $2.5 trillion and concentrated in a few key markets spanning over 80 countries. The *Union of Arab Banks' research division* has found

DOI: 10.1201/9781003462033-14

Table 12.1 Summary of fintech eras [1]

	Fintech 1.0	Fintech 2.0	Fintech 3.0	Fintech 3.5
Period	1866–1967	1967–2008	2008–present	Emerging
Geography	Global/developed	Global/developed	Developed	Emerging
Key elements	Infrastructure	Traditional financial digital services	Mobile/startups/new entrants	Mobile/startups/new entrants
Shift origin	Linkages	Digitalization	2008 financial crisis/ smartphone	Accessibility of new technology, particularly mobile communications in developing countries
Key events	• The first transatlantic cable (1866) • Fedwire (1918)	• The first ATM was installed by Barclays (1967) • Establishment of NASDAQ (1973) – the World's first stock exchange • SWIFT (Society for Worldwide Interbank Financial Telecommunications (1973) • The first Internet banking service was introduced by the Nationwide Building Society (1997) • The global financial crisis (2008)	• Bitcoin (2009) became the first digital decentralized currency • Google Wallet (2011) • Apple Pay (2014)	• The primary factor of recent Fintech developments in Asia and Africa has been the • need for economic growth • Alipay's introduction of facial recognition payment in March 2015 in China

that ten countries account for approximately 95% of the world's Shariah-compliant assets. Notably, Iran holds the top position with 29% of the market share, followed by Saudi Arabia (25%), Malaysia (11%), the United Arab Emirates (8%), Kuwait (6%), Qatar (6%), Turkey (2.6%), Bangladesh (2.1%), Indonesia (2%), and Bahrain (1.8%) [5].

These statistics demonstrate the significant presence of Islamic finance in specific regions, highlighting the concentration of Shariah-compliant assets in these countries. This data underscores the potential for Islamic Fintech to thrive in these markets, given their established foundations in Islamic finance and the growing demand for innovative financial solutions that align with Sharia principles.

The principles of Islamic finance are rooted in specific economic and commercial operations classified as permissible or prohibited under Sharia law, guided by its goals [6]. Islamic banking and finance are built upon the principle of risk-sharing. Understanding the role of risk-sharing in capital raising is crucial. Sharia prohibits Muslims from earning income through interest (usury) but allows income generation by sharing risks and rewards among the parties involved in a transaction. This profit-sharing mechanism encourages collaboration and partnership, rather than a traditional creditor–debtor relationship. Partnership fosters mutual responsibility for the outcome of financed projects, which is believed to enhance the likelihood of their success [7]. There are two primary methods of risk-sharing in Islamic finance:

Mudaraba: Also known as "silent partnership," Mudaraba involves all parties actively participating in the execution of the project. Investors provide the capital while the entrepreneurs manage the projects. Profits are distributed based on a predetermined profit-sharing ratio.

Musharakah: Often referred to as "equity partnership," Musharakah allows for shared ownership and profit-sharing. Investors and entrepreneurs jointly contribute capital and engage in the decision-making process. However, investors do not interfere in the day-to-day management of the project.

Both Mudaraba and Musharakah facilitate uncapped returns based on the agreed-upon profit-sharing ratio [8].

Islamic fintech provides a platform for unique and innovative startups to adapt and implement a risk-sharing model within Islamic financial institutions, thereby contributing to the growth of the Islamic finance sector. An example of this is the emergence of Islamic neo-banking or virtual banks, which offer prospects for financiers and business owners seeking Islamic risk-sharing and interest-free goods and services [9]. By incorporating the principles of risk-sharing, Islamic fintech promotes inclusivity, ethical practices, and alignment with Sharia law in the financial ecosystem.

Another primary objective of Islamic finance is to promote social and economic development through practices such as *Zakat*, a form of obligatory alms-giving, and other specific business principles. Most Islamic institutions have a board of religious advisers who provide guidance and judgment on the acceptability of new fintech processes [10]. These advisers ensure that new developments in Islamic fintech adhere to Sharia principles and contribute to the overall welfare and progress of society.

By integrating risk-sharing and emphasizing social and economic development, Islamic finance and fintech aim to create a financial system that aligns with Islamic values and promotes sustainable growth.

Prohibitions of Islamic finance

The distinction between conventional and Islamic banking lies in the utilization of specific techniques and concepts in traditional finance that are strictly prohibited by Sharia law. These prohibited practices include:

Investing in Businesses Involved in Forbidden Activities: Islamic principles prohibit engaging in activities such as the production and distribution of alcohol, pork, pornography, tobacco-related products, and weapons. These actions are considered *Haram* or unlawful. Therefore, investing in businesses involved in these prohibited activities is not permitted. The restriction extends beyond the direct buying and selling of prohibited goods and services to encompass the entire production and distribution chain, including packaging, transportation, warehousing, and marketing [11].

Paying or Charging Interest: Interest-bearing loans are regarded as Haram in Islam as they are seen to favor the lender at the expense of the borrower. Sharia law prohibits the charging or payment of interest, considering it as usury (Riba) [12]. *Riba*, which means "to increase" or "to exceed" in Arabic, refers to illegal transactions or interest charges imposed by lenders. Islamic banking operates on an interest-free basis, ensuring that all financial transactions and operations are free from interest [6].

Uncertainty and Risk: Islamic financial principles discourage participation in high-risk or uncertain contracts. The term *Gharar* refers to the concept of uncertainty or risk in investments. Examples of Gharar include short sales and derivative transactions, which are considered risky or speculative investments and are prohibited in Islamic finance [12]. Another instance of Gharar occurs when one party in a transaction possesses an undue advantage, creating information asymmetry. This lack of transparency and unequal distribution of knowledge is prohibited in Islamic finance.

Speculation: Sharia law strictly prohibits all forms of speculation or gambling, known as *Maisir*. Islamic financial institutions are forbidden from engaging in agreements where the ownership of goods depends on an uncertain future event [12]. Examples of Maisir include lotteries, bonds with prizes, futures, and options contracts [6].

By adhering to these prohibitions, Islamic banking ensures that financial practices align with the ethical principles of Islam. The focus is on promoting transparency, fairness, and risk-sharing while avoiding exploitative practices and uncertain transactions.

Alternative financing

Islamic banking and finance offer a distinct approach to financing that is rooted in Islamic Sharia law, principles, and regulations. Table 12.2 provides a concise overview of key Sharia-compliant financing arrangements.

In addition to the previously mentioned financing arrangements, there are several other contracts commonly used in Sharia-compliant finance:

Ijarah: An operating lease contract where the lessor leases an asset to the lessee for an agreed period in exchange for rental payments. This arrangement is commonly used for equipment, vehicles, or property leasing.

Istisna: A contract used for financing the construction, manufacturing, or production of goods or assets. Under this arrangement, the financier contracts with a manufacturer or contractor to build or produce a specific item according to agreed-upon specifications and payment terms.

Wadiah: A safe custody contract where a person deposits their assets or funds with another party who acts as a custodian. The custodian is responsible for safeguarding the assets or funds and returning them upon the depositor's request.

Wakalah: An agency contract where one party (the principal) appoints another party (the agent) to act on their behalf in a specific task or transaction. The agent is entrusted with certain responsibilities and acts in the best interest of the principal within the defined scope of authority.

Table 12.2 A Recap of Various Sharia-compliant Financing [13]

Mudaraba	*A partnership agreement where one side contributes funds while the other contributes managerial and entrepreneurial abilities to conduct a firm. The industrial or administrative partner (the "Mudarib") receives a predetermined portion of the profits, while the capitalist or general partner (the "Rabal-Maal") is responsible for any losses.*
Musharakah	A joint venture entails the ongoing equity investment of all parties to work on a specific project within a set time frame. A predetermined ratio is used to divide earnings and losses among all parties.
Murabaha	Purchasing products at a consumer's request, and the seller discloses the profit margin of the credit transaction. After buying the products from the supplier, the financial institution sells them to the client at a price that includes markup and a predetermined credit duration. The customer can pay a security deposit to the financial institution, and the remaining financing balance can be guaranteed or secured with collateral.

These additional contracts provide further flexibility in structuring Sharia-compliant financial transactions to meet the specific needs of individuals and businesses. They contribute to the diversity and adaptability of Islamic finance, allowing for a broader range of financing solutions in accordance with Sharia principles [11].

Blockchain and Islamic finance

Blockchain technology, with its decentralized and transparent nature, has the potential to revolutionize various aspects of Islamic finance. From a Sharia perspective, the use of technology, including Blockchain, is considered permissible as it acts as a neutral enabler. Blockchain operates by creating blocks that contain transaction records, timestamps, and cryptographic hash codes of preceding blocks. Its key features include resistance to data modification and the traceability of transactions.

Islamic finance can benefit from the monitoring capabilities provided by Blockchain. By utilizing smart contracts, financial transactions can be securely executed and recorded on the Blockchain. Smart contracts enable irreversible and traceable records of ownership and assets, reducing costs associated with financial arrangements and services significantly. For example, in crowdfunding platforms, smart contracts can ensure the secure recovery of pledged payments to donors, if funding criteria are not met by the project creators. This assurance can enhance the efficiency and trustworthiness of Islamic banking. To ensure the integration of Blockchain technology into Islamic banking and Sharia-compliant applications, Sharia governance is crucial. Specifically, the following areas should be considered:

Blockchain-Based Investments: A comprehensive Sharia evaluation of assets and Sharia-compliant companies should be conducted. If the activities facilitated by the Blockchain are incompatible with Sharia principles, support from Sharia-compliant banks should be withdrawn.

Businesses Adopting Blockchain: Evaluations should consider whether the Blockchain being developed is associated with illegal or unethical industries.

Blockchain Protocols: Sharia review should focus on the operation of algorithms, transaction costs, governance rules, and fair distribution of rewards. Transaction fees should be reasonable and fair. Incentives and rewards should be distributed in a Sharia-compliant manner, avoiding biased situations. Additionally, the rights of participants and nodes should align with Sharia's principles of justice and fairness. Specific terminologies used within the Blockchain system should also be examined to ensure compliance with Sharia principles.

By applying Sharia governance to Blockchain-based initiatives, Islamic finance can leverage the benefits of this technology while ensuring adherence to Islamic principles and values [14].

Cryptocurrency and Islamic finance

Cryptocurrencies, such as Bitcoin, have gained significant attention as digital currencies built on Blockchain technology. They serve as a medium of exchange and rely on the secure, traceable, and unchangeable nature of Blockchain for financial transactions [4]. While there are various digital currencies and tokens available, each with its own characteristics, such as higher rates, scalability, privacy, or the ability to establish Smart Contracts. The compatibility of cryptocurrencies with Islamic banking principles has been a topic of discussion among Islamic scholars, financial experts, and bankers [15].

Many Islamic scholars have expressed concerns about the compliance of cryptocurrencies, including Bitcoin with Islamic principles. They have issued *Fatwas* declaring cryptocurrencies as Haram, meaning prohibited under Islamic law. In Islam, a fatwa is a legal ruling or interpretation given by an Islamic scholar or mufti on a specific issue based on their understanding of Sharia. Fatwas can cover a wide range of topics, including religious practices, personal conduct, social issues, financial matters, and more. The reasoning behind these Fatwas is that cryptocurrencies lack legal tender status, have unidentified issuers, operate without a central authority, exhibit fluctuating values, and can be easily exploited for illegal activities [16]. Consequently, Bitcoin is generally regarded as non-compliant with Sharia and should be prohibited [4].

However, as the popularity and demand for Bitcoin continue to rise, some academics argue that it represents a groundbreaking invention deserving recognition. They contend that the high price of Bitcoin does not necessarily imply a violation of Sharia law but rather reflects increasing acceptance and belief in cryptocurrencies. These scholars suggest that Bitcoin can be analyzed in terms of risk-sharing instead of risk shifting. They argue that millions of Muslim adults benefit from Bitcoin and other cryptocurrencies through risk-sharing and *Maslaha* (benefits to society), which indicate a potential positive aspect. Bitcoin operates based on collaboration and shared risk among multiple participants, leading to risk reduction through a large user pool. Evaluating cryptocurrencies on a transaction-by-transaction basis to determine their compliance with Sharia is a challenging task, making it difficult to universally categorize all cryptocurrencies as forbidden. There is potential to leverage Blockchain technology to create a Sharia-compliant digital currency, although this would require modifications to the current circulation of cryptocurrencies [4].

Within the Islamic scholarly community, there exists a range of opinions regarding the permissibility (*Halal*) or impermissibility (*Haram*) of cryptocurrencies. The lack of clear guidance in Islamic law on the use of cryptocurrencies in transactions has led to the need for a concept and system for Islamic digital currency. Various arguments support categorizing cryptocurrencies as prohibited, including constitutional violations highlighted by several Fatwas and researchers. Another reason to consider cryptocurrencies as illegal is their susceptibility to value fluctuations resulting from factors such as system hacking and technical issues. While cryptocurrencies like Bitcoin can be deemed acceptable in certain situations, such as for currency exchange or payment for goods and services, their purchase for the purpose of accumulation or investment is prohibited due to their high volatility in exchange value [17]. The manner in which a cryptocurrency is acquired also plays a crucial role in determining its

Table 12.3 Permissibility/impermissibility factors as per the sharia perspective [4]

Impermissible (Haram)	Permissible (Halal)
Due to constitutional violations and implementation problems with cryptocurrencies, it is impermissible.	Based on its guiding principles and nature, cryptocurrency can be permissible.
It is not permitted to purchase them with the intent to accumulate or invest in them.	Permissible in some circumstances, such as exchanging currencies or paying for products or services.

permissibility. If a cryptocurrency is purchased specifically for settling payments in exchange for goods or services, it is considered acceptable. However, if the cryptocurrency is acquired through mining with the expectation of obtaining significant future returns, it is deemed to involve excessive risk (Gharar) and is therefore forbidden [4].

The lack of comprehensive guidance in Islamic law has led to ongoing debates and discussions on the topic. It is important to consider the specific circumstances and purposes for which cryptocurrencies are used to determine their compliance with Islamic principles (Table 12.3).

Potential use cases in Islamic finance

Remittance: Islamic banks worldwide are adopting digitization of banking services, including peer-to-peer (P2P), business-to-business (B2B), business-to-individual (B2I), and e-commerce transactions to improve the accuracy, security, speed of processing, and lower processing costs for remittance. Blockchain technology can enable users to send payments over the Internet, facilitating more effective wealth transfer through cross-border digital transactions [14]. By utilizing decentralized databases and peer-to-peer networks, Blockchain allows for the digitization of data, ensuring its security and protection against tampering or deletion. This unique feature of Blockchain technology eliminates uncertainty (Gharar) and enables affordable transaction processing, promoting financial inclusion for unbanked migrant workers and organizations following Islamic Finance law that need to send money abroad.

Sukuk: Also known as Islamic or Sharia-compliant bonds, Sukuk represents a percentage of ownership in a portfolio of qualifying assets. They offer an Islamic variation of traditional bonds [14]. Different Sukuk structures are built on various Sharia-compliant contracts, such as profit-sharing, deferred-delivery purchase, leasing of assets, joint ventures, project-based contracts, and cost-plus asset purchase. Sukuk has been widely used by governments to finance infrastructure projects but has been less accessible to smaller businesses due to its complexity and high issuance costs. Blossom Finance introduced the "Smart Sukuk" platform, leveraging Ethereum Blockchain smart contracts to improve the efficiency and reach of Sukuk issuance globally. This platform aims to standardize and automate the accounting, legal, and administrative aspects of Sukuk offerings, enabling small and medium-sized businesses, social impact projects, groups, and associations, to issue their Sukuk using this new technology [9]. By embedding business rules in smart contracts on the Ethereum Blockchain, organizations can raise funds by offering tokens representing a share of the Sukuk's ownership, eliminating the need for traditional banks or intermediaries [18].

Waqf: A charitable donation made permanently and committed to *Allah*. It can be in the form of fixed assets or cash [18]. To ensure good governance and transparency in managing Waqf funds, Blockchain technology can be employed. By utilizing a shared and immutable ledger, Blockchain provides real-time, transparent data accessible to authorized members of a permissioned network. Blockchain networks can keep track of Waqf contributions, investments, allocations, and cash flows, allowing stakeholders to review every step of a transaction, enhancing confidence and visibility across the entire Waqf value chain.

Takaful: Also known as Islamic insurance, Takaful operates under principles that prohibit compensation for harm or loss to policyholders. The uncertainty (Gharar) component found in conventional insurance contracts is not permitted in Takaful. Instead, Takaful employs a "donation contract" among participants, where a predetermined portion of contributions is allocated to help other participants in the event of a loss. Introducing Blockchain in Takaful improves data security and confidence among parties involved. By utilizing shared and immutable ledgers, Blockchain enables the use of smart contracts to efficiently manage and process claims, detect and prevent fraud, and verify the authenticity and ownership of items and documents [14].

Smart Contracts: These self-executing contracts are designed to automate processes, verification, and execution without the need for intermediaries. They reduce uncertainty (Gharar) by verifying contractual terms only when conditions are met. Smart contracts can automate the entire contractual process, mitigating operational risks and reducing redundancies and complications. In Islamic banking, smart contracts can aid in avoiding significantly rising interest rates, minimizing speculation and uncertainty [18].

Zakat Collection: An obligatory charity requirement in Islam. Muslims must pay a specified sum of money or property to those who qualify to receive it. The *Nisab*, or basic standard of living index, determines the minimum threshold for paying Zakat. Assets exceeding this threshold by individuals for a whole year are subject to Zakat, calculated at 2.5% of all liquid assets held for at least a year. With the inclusion of cryptocurrencies, Muslims must carefully account for their holdings in the Zakat calculation. Blockchain technology can make the Zakat process auditable, immutable, and trackable, ensuring transparency and enabling quick identification of any potential flaws or errors [18].

Challenges of implementing blockchain in Islamic finance

Security Challenges: Similar to most other blockchain use cases, implementing blockchain technology in Islamic finance introduces security risks such as data loss, hacking, identity theft, and online fraud. One specific concern is the majority or 51% attack, where a party gains control of over 50% of the hashing power on a public blockchain. This enables them to manipulate the blockchain, halt transactions, and potentially engage in double-spending. The *Ghash.io* Incident in 2014 exemplifies the severity of these threats [19].

Challenges related to Sharia Compliance: Sharia compliance is vital for Islamic financial institutions, requiring any new application to adhere to Sharia-compliant standards. The absence of such standards in blockchain technology poses a challenge. Furthermore, the correct sequencing, adherence to approval conditions, and implementation of smart contracts are essential to ensure compliance with Islamic financial regulations. Islamic financial institutions must verify the Sharia compliance of any new technology or mechanism employed [19].

Technology Infrastructure Challenges: The shortage of qualified talent has been a significant issue in Islamic banking and finance since its inception. Islamic financial institutions have faced a shortage of professionals skilled in Sharia, Takaful, and other fields, impacting the industry's growth. Additional challenges arise due to the evolving nature of fintech, which demands expertise in information technology, data science, programming, and Sharia. This combination of skills makes talent acquisition more complex. As such, leaders of Islamic Finance Institutions (IFIs) should possess knowledge of new technologies, business, economics, finance, and Sharia [9].

Furthermore, incorporating blockchain technology into the existing infrastructure of financial institutions incurs high costs, including the operational costs of the blockchain itself and mining operations. The electricity required for continuous mining operations and the cost of computers used in the mining process contribute significantly to these expenses. Moreover, financial institutions may face challenges in abandoning existing programs and tools, which have already incurred substantial costs. Thus, adopting a new mechanism poses a significant challenge for financial institutions [19].

> *Regulatory Issues*: Regulating Islamic fintech presents difficulties due to its rapid development and potential for disruptive innovation. Policymakers and regulators face a dual challenge of harnessing the benefits and managing the associated risks. Adopting a permissive, liberal, and principle-based regulatory approach can facilitate innovation while ensuring proper conduct and compliance in the fintech industry. Establishing a regulatory framework based on principles that promote business stability, sustainability, and easy access to financial services is crucial. However, the decentralized nature of fintech services complicates the implementation of consistent regulatory rules across all sectors [4].

Regulatory support is essential for creating a favorable economic environment for business owners, entrepreneurs, and Islamic financial institutions. Some Islamic nations, like Malaysia, have successfully established regulatory climates conducive to innovation in the fintech sector, including the Islamic ecosystem [9].

A proposed Ethereum-based platform model

Ethereum is an open-source, blockchain-based computing platform and operating system that enables the creation of decentralized applications (DApps) and smart contracts. The acceptance and participation of the Islamic community in Ethereum, particularly in the development of smart contracts and DApps, as well as in mining and trading Ether (ETH), have raised questions regarding Sharia compliance [20].

Ether serves as the utility token and native cryptocurrency of the Ethereum platform, allowing users to execute smart contract operations. In the context of Sharia, Ether is considered an asset or wealth *(Mal)* due to its unique value and purpose within the Ethereum ecosystem. *Mal* refers to anything that can be obtained and possessed, whether in physical or digital form. Ether, being a digital asset that is transparent and secure on the public Ethereum Blockchain, does not possess any characteristics prohibited in Islam [21].

Amanie Advisors, an Islamic finance advisory firm, has stated that Ether is necessary for the operation of the Ethereum Blockchain as a trading tool. Introducing a reward system like Ether is crucial for creating a decentralized platform and harnessing its disruptive potential. Therefore, Ether is considered a Sharia-compliant asset (Mal), enabling Muslims to legitimately trade, exchange, and hold Ether to fully participate in the Ethereum ecosystem [20].

Islamic Coin is a digital currency designed to comply with Sharia principles and cater to the global Muslim community. It operates on *Haqq*, an Islamic Blockchain that strictly adheres to Islamic financial principles and customs. Haqq is a proof-of-stake Blockchain network compatible with Ethereum and offers instant resolution. By leveraging the existing codebase and toolkit of Ethereum, developers can build applications on Haqq without the need to rewrite existing smart contracts. Haqq holders can actively participate in network consensus and receive incentives for network protection through the proof-of-stake method [22].

In summary, Ether is considered Sharia-compliant as a tradable asset within the Ethereum ecosystem, allowing Muslims to engage in Ethereum-based activities. Additionally, the development of Islamic Coin and Haqq provides further opportunities for the Muslim community to participate in blockchain technology while adhering to Islamic financial principles.

CONCLUSIONS AND RECOMMENDATIONS

This chapter includes a review of Islamic finance and some blockchain implementation considerations in Islamic finance, along with its challenges. Islamic Fintech has a promising future in the Islamic World and among Muslim consumers of financial services since it offers innovation opportunities and can deliver financial services at a reasonable price. Islamic Fintech, or the use of Fintech in Islamic finance, explores many prospects while posing many problems. Islamic Fintech can help firms grow because it is transparent, simple to use, and accessible. It also quickly builds client confidence, which is crucial for startups. Compared to conventional finance and banking, fintech solutions are more cost-effective for providing financial services. As Islamic fintech is still in its infancy, several challenges come with great opportunities, including a lack of excellent and authentic research in the field, a lack of trained human staff, trade-offs between government and Sharia compliance, cyberattacks, and investor confidence. Another crucial issue for Islamic Fintech is the regulatory environment, which needs to be adhered to. When policing the Fintech industry, the regulatory bodies should consider more permissive, liberal albeit principle-based approaches.

For Islamic Fintech, Blockchain technology offers a more innovative and secure business methodology. Using Blockchain technology, transactions are more transparent and accessible to all users. The monitoring and regulatory process can be simplified to the simple act of establishing a smart contract, and smart contracts can be a beneficial mechanism in all financial transactions. Cryptocurrencies have been a revelation, and Muslim nations urgently need to conduct further research and figure out how to create a cryptocurrency that is entirely compliant with Sharia law. Additionally, Islamic finance can be conceptualized as "interest-free banking" in a limited sense, which contrasts with the traditional interest-based banking system people typically deal with. The fundamental ideas of Islamic finance focus on the types of economic and commercial activity that Sharia and its goals prohibit and permit (*Maqsid*). Some distinctive characteristics of the Islamic financial system include the prohibition of transactions involving the payment of interest (*Riba*), speculation (*Maisir*), uncertainty (*Gharar*), and the approval of profits from genuine transactions.

According to Amanie Advisors and the Ethereum Foundation, Ether, the native cryptocurrency of the Ethereum network, satisfies the requirements to be classified as an asset or Mal from a Sharia perspective. Yet, one of the critical conditions for ensuring that the Blockchain works as intended is that it incorporates a medium of exchange. Ether is a Sharia-compatible asset since it lacks banned elements like Riba, Gharar, and Maysir. Therefore, greater involvement from the Islamic financial sector and the Muslim community to fully utilize the Blockchain technology provided by the Ethereum platform for creating various Sharia-compliant decentralized applications and smart contracts can be planned and implemented.

Good practices for using blockchain technology in Islamic finance

Blockchain technology has the potential to revolutionize the Islamic finance sector, but regulations must be followed to ensure its effective implementation. The following is a recap of some recommended good practices for the effective implementation of Blockchain technology in Islamic finance.

- *Adhere to Sharia Principles*: Blockchain technology used in Islamic finance must comply with Sharia principles, which prohibit Riba (usury), uncertainty in contracts (Gharar), speculation (Maysir), fraud, bribery, illegally taking property from others, and dealing in forbidden (Haram) items.
- *Conduct Thorough Research*: Before implementing Blockchain technology in Islamic finance, conduct comprehensive research to ensure that the technology is compliant with Sharia principles such as fairness, transparency, and risk-sharing and to identify potential challenges.
- *Train Staff*: Train staff to understand the nuances of Blockchain technology and Islamic finance to ensure they can efficiently implement the technology. To effectively identify, define, select, implement, and improve benchmarking for the training programs made available to Islamic finance practitioners requires the necessary activities and tasks to understand Blockchain technology.
- *Use of Smart Contract*: The use of smart contracts is beneficial for institutions. it is essential to review the mechanism of the contract arrangement and ensure its correct sequence adherence, adherence to approval conditions, and subsequent implementation among other matters that ensure compliance concerning the legal side in Islamic financial dealings.
- *Maintain Adequate Regulatory Compliance*: Regulatory compliance is a key consideration for any Blockchain-based solution in Islamic finance. Sharia law has specific requirements for financial transactions and any Blockchain-based solution must be compliant with these requirements to be viable.
- *Use Permissive and Principle-based Approaches*: Regulatory bodies should adopt permissive and principle-based approaches when policing the Fintech industry. This approach will encourage innovation and growth while ensuring that the technology is compliant with Sharia principles.
- *Collaborate with Ethereum*: The Ethereum platform offers Blockchain technology that is Sharia-compliant and can be used to build a variety of Sharia-compliant decentralized applications and smart contracts. To completely utilize this technology, greater planning and implementation will need to be done with the help of the Islamic financial sector and the Muslim community.

REFERENCES

1. Arner, D. W., Barberis, J. N., & Buckley, R. P. (2015, October 20). *The evolution of Fintech: A new post-Crisis paradigm?* SSRN. Retrieved October 16, 2022, from https://papers.ssrn.com/sol3/papers.cfm?abstract_id=2676553
2. Guardiola Agustí, G. (2022, June 1). *Fintech the new challenge for banking.* Repositori Obert UdL. Retrieved November 9, 2022, from https://repositori.udl.cat/handle/10459.1/83541
3. Sharma, I. (2022, November 14). *What is Fintech and why is it important? - tatvasoft blog.* TatvaSoft. Retrieved March 19, 2023, from https://www.tatvasoft.com/outsourcing/2021/04/what-is-fintech.html#why-fintech
4. Rabbani, M. R., Khan, S., & Thalassinos, E. I. (1970, January 1). *Fintech, Blockchain and Islamic Finance: An extensive literature review.* L-Università ta' Malta. Retrieved October 7, 2022, from https://www.um.edu.mt/library/oar/handle/123456789/54860
5. Domat, C. (2020, November 5). *Islamic Finance: Just for Muslim-majority nations?* Global Finance Magazine. Retrieved January 24, 2023, from https://www.gfmag.com/topics/blogs/islamic-finance-just-muslim-majority-nations

6. Obana, J. (2018, March 9). *Prohibitions and alternative financing in Islamic Finance.* LinkedIn. Retrieved November 11, 2022, from https://www.linkedin.com/pulse/prohibitions-alternative-financing-islamic-finance-obana-cpa-mba/

7. Tabash, M., & Dhankar, R. (2014, February). *The relevance of Islamic Finance Principles in economic growth.* Retrieved March 19, 2023, from https://www.researchgate.net/profile/Mosab-Tabash/publication/327201573_The_Relevance_of_Islamic_Finance_Principles_in_Economic_Growth/links/5b7fd72892851c1e122ebf29/The-Relevance-of-Islamic-Finance-Principles-in-Economic-Growth.pdf?origin=publication_detail

8. Bakar, I.-F. A. (2019, March 19). *Risk-sharing or risk-transfer: Which is more relevant to Islamic banking?* LinkedIn. Retrieved March 19, 2023, from https://www.linkedin.com/pulse/risk-sharing-risk-transfer-which-more-relevant-abu-bakar/

9. Mohamed, H., & Ali, H. (2022). *Blockchain, Fintech, and Islamic Finance Building the future in the new Islamic Digital Economy.* De Gruyter.

10. Kamdzhalov, M. (2020, March 16). *Islamic Finance and the New Technology Challenges.* View of Islamic finance and the new technology challenges. Retrieved October 19, 2022, from https://www.ojs.unito.it/index.php/EJIF/article/view/3813/pdf

11. Moinuddin, C. (2015, March). *An introduction to Islamic finance – chartered institute of management …* Retrieved October 17, 2022, from https://www.cimaglobal.com/Documents/Islamic%20finance/Rebrand%20Brochures/Islamic%20Introduction%20brochure_Mar2015.pdf

12. *Islamic Finance.* Corporate Finance Institute. (2022, May 7). Retrieved October 16, 2022, from https://corporatefinanceinstitute.com/resources/knowledge/finance/islamic-finance/?gclid=Cj0KCQjw166aBhDEARIsAMEyZh7loO3GvLF1XOW1250KEjt7jZSb9NARK71bVdnWSaykpvc6lcxNLywaAiO9EALw_wcB

13. Alawode, A. (2015, March 31). *Islamic Finance.* World Bank. Retrieved January 24, 2023, from https://www.worldbank.org/en/topic/financialsector/brief/islamic-finance

14. Adam, M. F. (2022, March 6). *Integrating Blockchain in Islamic Finance.* Amanah Advisors. Retrieved October 7, 2022, from https://amanahadvisors.com/integrating-Blockchain-in-islamic-finance/

15. Muedini, F. (2018, August). *View of the compatibility of Cryptocurrencies and Islamic Finance.* Retrieved February 4, 2023, from https://www.ojs.unito.it/index.php/EJIF/article/view/2569/pdf

16. Selcuk, M., & Kaya, S. (2021, January 15). *Issues.* Turkish Journal of Islamic Economics. Retrieved February 4, 2023, from https://www.tujise.org/issues

17. Abu-Bakar, M. (2017, April 5). *Sharia analysis of bitcoin, Cryptocurrency, and Blockchain.* Retrieved February 5, 2023, from https://islamicbankers.files.wordpress.com/2019/02/2017-Sharia-analysis-of-bitcoin-cryptocurrency-Blockchain.pdf

18. Elasrag, H. (2019, March 6). *Blockchains for Islamic Finance: Obstacles & Challenges.* Munich Personal RePEc Archive. Retrieved October 7, 2022, from https://mpra.ub.uni-muenchen.de/92676/

19. Alaeddin, O., Al Dakash, M., & Azrak, T. (2021, July 1). Implementing the Blockchain technology in Islamic financial industry: Opportunities and challenges. *Journal of Information Technology Management.* Retrieved November 15, 2022, from https://jitm.ut.ac.ir/article_8326.html

20. *Sharia white paper on Ether Amanie Advisors Ethereum Foundation.* (2019, August). Retrieved March 6, 2023, from https://amanieadvisors.com/v4/wp-content/uploads/2019/08/Sharia-Whitepaper-Ether-v1.0.pdf

21. Gabriel, B. (2019, November 7). *Ethereum deemed halal by Muslim scholars, may stimulate ETH demand.* CryptoSlate. Retrieved March 5, 2023, from https://cryptoslate.com/ethereum-halal-muslim-scholars-eth-demand/

22. *Addressing talent scarcity in Islamic Finance.* Finance Accreditation Agency (FAA) (1012469-W). (n.d.). Retrieved March 20, 2023, from https://www.faa.org.my/article/addressing-talent-scarcity-in-islamic-finance

Index

Printed in the United States
by Baker & Taylor Publisher Services